Lecture Notes in Physics

Edited by J. Ehlers, München, K. Hepp, Zürich, H. A. Weidenmüller, Heidelberg, and J. Zittartz, Köln
Managing Editor: W. Beiglböck, Heidelberg

55

Nuclear Optical Model Potential

Proceedings of the Meeting
Held in Pavia, April 8 and 9, 1976

Edited by S. Boffi and G. Passatore

Springer-Verlag
Berlin · Heidelberg · New York 1976

Editors

S. Boffi
Istituto di Fisica Teorica
Università di Pavia
Via Bassi, 4
27100 Pavia/Italy

G. Passatore
Istituto di Scienze Fisiche
Università di Genova
Via Benedetto XV, 5
16132 Genova/Italy

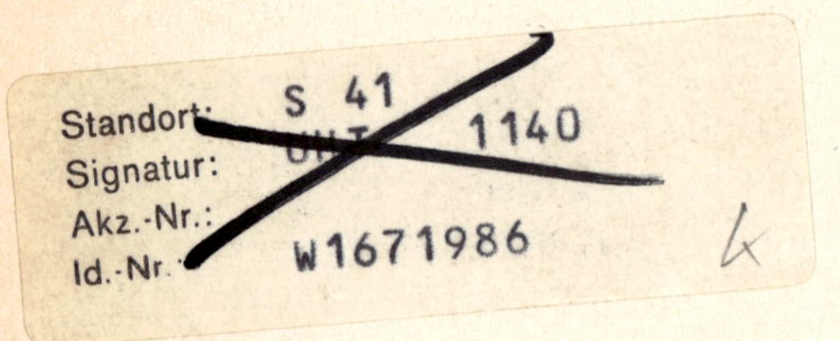

ISBN 3-540-07864-9 Springer-Verlag Berlin Heidelberg New York
ISBN 0-387-07864-9 Springer-Verlag New York Heidelberg Berlin

Printing and binding: Beltz Offsetdruck, Hemsbach/Bergstr.

PREFACE

The interaction of a nucleon of moderate energy with a nucleus is an exceedingly complicated process. However, confining the discussion to elastic scattering, it has long been recognized that an adequate description can be achieved by means of a single particle potential varying smoothly with energy and nuclear size. In principle, this potential should be non-local and energy dependent. Experimental and theoretical efforts have shown how to parametrize simpler forms of this potential in order to reproduce positions, widths and wave functions of single particle states in nuclei. This phenomenological approach has already given satisfactory results. In particular, the recent folding model calculations of the optical potential can be extended to include scattering of alpha-particles and heavy ions provided accurate nuclear density distributions are available.

The progress both of theoretical investigations and of measurements of fine details in the cross section and the polarization of nucleons makes this the right moment to reconsider the whole problem of the optical model potential in a unified approach, including also bound states.

The very recent applications of nuclear matter techniques to calculate the mass operator in terms of realistic nucleon-nucleon interactions, made in the last two years, constitute a very important theoretical achievement. They provide a direct evaluation of the various properties of the theoretical optical potential. In particular a comparison is now possible with the non-locality and the energy dependence already derived by dispersion relation techniques in a rather indirect way but without involving any specific nucleon-nucleon interaction. The substantial agreement between the results of such different approaches, as well as with that derived in the multiple scattering formulation, is impressive and has not been pointed out earlier, since some of these results are very recent.

Furthermore, in the last years phenomenological non-local models, indicated by the various theoretical formulations, have been improved and comparison of their implications with recent experimental data gives significant information.

It seems well indicated that the different pieces of information coming from different sources will produce a satisfactory consistent picture. This volume presents the proceedings of a small symposium held for this purpose in Pavia, April 8 and 9, 1976, within the frame of a joint activity of the Faculties of Mathematical, Physical and Natural Sciences of the Universities of Genova and Pavia, Italy, and sponsored by the Istituto Nazionale di Fisica Nucleare to whom we would like to express our gratitude.

o o o

The first paper (by G. Passatore) has an introductory character and gives a survey of the various derivations of the theoretical optical potential for scattering processes, and the second one (by F. Capuzzi) deals in particular with the problem of defining a local potential equivalent to a non-local potential. After an introduction on the use of Green functions, an application is presented (by S. Boffi) to discuss some recent experimental results by knock-out reactions.

The first paper by J.-P. Jeukenne, A. Lejeune and C. Mahaux reports the very recent calculation of the theoretical optical potential by the nuclear matter techniques. The folding model calculation for the optical model potential of alpha-particles and heavy ions is then presented by P. E. Hodgson, and the relevance of the accurate knowledge of the nuclear density distributions is pointed out. The following paper (by M. Giannini and G. Ricco) exhibits a non-local energy independent optical model suggested by the multiple scattering approach and its use in fitting the experimental data on several nuclei. In a parallel phenomenological approach the short paper by F. Malaguti reports on the systematics of the $1f_{7/2}$ neutron single particle energies. Some very recent experimental results are presented by a group in Pavia (G. Bendiscioli et al.) on scattering of polarized protons on several nuclei at very small angles (below 7°). Various treatments of the energy dependence of the optical potential by dispersion relation techniques are discussed by G. Passatore and compared with those based on the mass operator method and on the multiple scattering approach. Finally, the last paper (by J.-P. Jeukenne, A. Lejeune and C. Mahaux) gives a survey and an analysis of the present theoretical and phenomenological situation.

S. Boffi and G. Passatore

TABLE OF CONTENTS

A SKETCH OF THE VARIOUS FORMULATIONS OF THE THEORETICAL OPTICAL POTENTIAL FOR SCATTERING PROCESSES

G.PASSATORE

Istituto di Scienze Fisiche dell'Università di Genova

Istituto Nazionale di Fisica Nucleare,Sezione di Genova

Abstract. The various formal derivations of the theoretical optical potential for scattering problems are outlined and their connections are discussed. Such a survey concerns the time dependent derivations as well as the stationary formulations in terms of the spectral properties of the many body system, of the nucleon-nucleon scattering amplitude and of the mass operator. The different pieces of information that every formulation gives are pointed out and the situations where each representation of theoretical optical potential is more suitable to obtain particular results are discussed. A brief comment on the mathematical and formal attitudes to handle this problem is also made.

1. Introduction

Often, when one speaks of "optical potential", some confusion may arise.

A first reason is that theoretical and experimental physicists speak about things which are rather different, although if stricIty related; a second one lies in the fact that the theoreticians derive the optical potential in various different ways.

I shall call the "theoretical optical potential" (TOP) that derived on a theoretical ground, whichever it may be, and the "empirical optical potential" the one employed in the phenomenological analyses.

In the following of this paper I shall briefly sketch several derivations of the TOP for scattering processes in order to propose a uniform notation for this meeting, to put in evidence their connections and to discuss the situations where each formulation is more suitable to be used.

Section 2 deals with the approaches based on the many-channel scattering theory. The time-dependent formulation and the stationary derivations in terms of the spectral properties of the manv-body system and of the nucleon-nucleon scattering amplitudes are outlined and compared. Section 3 deals with the identification of the theoretical optical potential with the mass operator.

The next paper, dealing with the local equivalent potential, will also cover the point of the connection between the theoretical optical potential and the empirical optical potential.

2. Derivations based on the many-channel scattering theory

There is a variety of approaches, according to their starting from the time dependent Schrödinger equation, the stationary Schrödinger equation, the Lippmann-Schwinger equations for the wave function or for the T-matrix.

The common idea is to project the equation for the state vector (or for the T-matrix) of the whole system, which describes all the types of processes, into the "elastic channel subspace", which is the subspace formed by the state vectors describing the state of the many-body system with the nucleus in its ground state. Such a projected equation is by itself of a Schrödinger type, or, respectively, of a Lippmann-Schwinger type, with an operator which plays the role that the ordinary potential plays in the potential scattering theory. This is still a many-body operator. After reducing this equation to a two body equation for the relative motion projectile-nucleus, such an operator becomes the TOP.

In such formalisms a fundamental role is played by the projection operator P on the elastic channel subspace.
As long as the identity between the projectile and the target nucleons is neglected, this operator is simply:

$$P = |0\rangle\langle 0| \quad , \tag{1}$$

where $|0\rangle$ is the state vector of the correlated ground state of the nucleus. We introduce also the projection operator on all the anelastic channels:

$$Q = 1 - P. \tag{2}$$

A more sophisticated expression of the operator P is required if the antisymmetrization between the projectile and the target nucleons must be taken into account: such an expression has been derived by Feshbach. [1)]

In the following comparison of the various formulations of the TOP we are going to use the projection operator in the form valid for distinguishable particles. At the end we shall add some comments on the problem of the antisymmetrization in the various formulations.

2.1. The time-dependent formulation

The formal derivation of the TOP starting from the time dependent Schrödinger equation has been made by Coester and Kummel.[2)] The steps are as follows. From the time dependent Schrödinger equation for the state vector $|\Psi(t)\rangle$ of the many-body system, one obtains two coupled equations for the components $P|\Psi(t)\rangle$ and $Q|\Psi(t)\rangle$. Both equations involve the first time derivative, and so the elimination of $Q|\Psi(t)\rangle$ implies an integration. Therefore one gets for $P|\Psi(t)\rangle$ an integro-differential equation. After eliminating the nuclear variables by taking the scalar product with $|0\rangle$, one obtains the equation for the vector $\langle 0|P|\Psi(t)\rangle$ which describes the relative motion particle-nucleus:[3)]

$$(3)\qquad i\hbar(d/dt)\langle 0|P\Psi(t)\rangle = (H_0+\langle 0|V|0\rangle)\langle 0|P\Psi(t)\rangle + \int_{-\infty}^{t} U(t-z)\langle 0|P\Psi(z)\rangle\, dz \quad ,$$

where:

H_0 is the kinetic energy of the projectile and the c.m.of the target;

V is the interaction between the projectile and the target;

$$(4)\qquad U(t) = -i\,\langle 0|PVQ\exp\left(-(i/\hbar)tQHQ\right)QVP|0\rangle \quad .$$

It must be emphasized that Eq. (3) contains the boundary condition for $t\to-\infty$, i.e. that the process originates in the elastic channel. This is a very important feature, which in the following will be called "causality condition". It can also be stated in the form:

$$(5)\qquad U(t) = 0 \ , \text{ for } t<0 \quad .$$

Of course, the integral in Eq. (3) is a strong limit and it may be replaced by the Abelian limit:[3)]

$$(6)\qquad \underset{T\to-\infty}{\text{s-lim}} \int_{T}^{t} U(t-z)\langle 0|P\Psi(z)\rangle dz = \underset{\eta\to 0^+}{\text{s-lim}} \int_{-\infty}^{t} U(t-z)\exp\left(-\eta(t-z)\right)\langle 0|P\Psi(z)\rangle dz \ .$$

In the r.h.s. of this equation the causality condition is expressed by the infinitesimal positive quantity η. Remembering the condition (5) for $U(t)$, Eq. (3) can be written as

$$(7)\qquad -i\hbar(d/dt)\langle 0|P\Psi(t)\rangle=\left(H_0+\langle 0|V|0\rangle\right)\langle 0|P\Psi(t)\rangle + \underset{\eta\to 0+}{\text{s-lim}} \int_{-\infty}^{+\infty} U(t-z)\exp\left(-\eta(t-z)\right)\langle 0|P\Psi(z)\rangle dz \ .$$

2.2. The stationary formulation

2.2.1 The formulation starting from Schrödinger equation

Feshbach's formulation [1,4] starts from the stationary Schrödinger equation for the many-channel wave function Ψ_E , with the boundary condition that in the elastic channel both incoming and outgoing waves are present, but all the anelastic channels contain only outgoing waves. This is the stationary form of the causality condition presented above. One then obtains a set of coupled equations for $P\Psi_E$ and $Q\Psi_E$. In order to eliminate $Q\Psi_E$, one needs the resolvent of the operator QHQ, and the outgoing wave condition imposed on the anelastic component $Q\Psi_E$ requires that such a resolvent be defined as

$$(8) \qquad (E-QHQ+i\eta)^{-1} \ , \quad \eta\to 0^+ \ ,$$

for E above the threshold of the continuous spectrum of QHQ.

The equation for the amplitude $\langle 0|P\Psi_E\rangle$ obtained in this way is the following:

$$(9) \qquad \left(H_0+\langle 0|V|0\rangle+\langle 0|VQ\ (E-QHQ + i\eta)^{-1}\ QV|0\rangle\right) \langle 0|P\Psi_E\rangle = E\langle 0|P\Psi_E\rangle \ .$$

The operator

$$(10) \qquad \langle 0|VQ\ (E-QHQ + i\eta)^{-1}\ QV|0\rangle$$

is just the Fourier transform of the operator U(t) given by Eq. (4). Of course, the causality condition on the operator U(t) reflects itself on the analyticity properties of the operator (10), which must be analytical in the upper half of the complex E-plane.

The operator

$$(11) \qquad \mathcal{V}=\langle 0|V|0\rangle+\langle 0|VQ\ (E-QHQ + i\eta)^{-1}\ QV|0\rangle$$

is the theoretical optical potential. The properties of the many body system are embodied in the operator QHQ, which becomes the protagonist in this formulation.

By introducing its spectral representation:

(12) $$QHQ = \sum_n E_n |\Phi_n\rangle\langle\Phi_n| + \int_{\varepsilon}^{+\infty} dE' \int d\alpha \; E' |\Phi_{E'\alpha}\rangle\langle\Phi_{E'\alpha}| ,$$

where

(13) $$QHQ|\Phi_n\rangle = E_n|\Phi_n\rangle , \quad QHQ|\Phi_{E,\alpha}\rangle = EQHQ|\Phi_{E,\alpha}\rangle ,$$

the theoretical optical potential can be written as:

(14) $$\mathcal{V} = \langle 0|V|0\rangle + \sum_n \frac{\langle 0|VQ|\Phi_n\rangle\langle\Phi_n|QV|0\rangle}{E_n - E} + \int_{\varepsilon}^{+\infty} \frac{dE'}{E-E'+i\eta} \int d\alpha \langle 0|VQ|\Phi_{E'\alpha}\rangle\langle\Phi_{E'\alpha}|QV|0\rangle .$$

Such an expression can be also derived in the framework of the Wigner R-matrix theory[5] of nuclear reactions as well as in the Kapur-Peierls formalism.[6]

Some properties are worthwhile being remarked:

1) Analiticity property: it has been already remarked. In Eq. (11) the limit of the operator $\mathcal{V}$ for E going to the real axis from above must be used.

2) Spectral property: the spectrum of the operator QHQ determines the singularities of $\mathcal{V}$ on the real axis. These consist in poles E_n and in a cut starting from the beginning of the continuous spectrum of QHQ, which consequently takes on the meaning of the threshold for the anelastic processes.The non hermitian part of $\mathcal{V}$, which accounts for the lack of the continuity equation in the elastic channel, starts from this point.

3) Asymptotic behaviour with the energy: it would be very desirable to get from the representation (15) some information on the asymp-

totic behaviour. But one gets only the indication that this depends on the operator

$$\int d\alpha \langle O|VQ|\Phi_{E,\alpha}\rangle\langle\Phi_{E,\alpha}|QV|O\rangle$$

which, if the theoretical optical potential is considered from a relativistic point of view, must also contain the contribution from all the channels of particle production which are opened by the operator V which becomes a relativistic field-theory operator.

The relevance of the representation (14) of the theoretical optical potential is twofold:

1) It is suitable to describe the resonances in the nuclear scattering. When only one pole dominates, the T-matrix corresponding to the potential (14) consists of a Breit-Wigner term plus a background term, and so it gives the usual parametrization of the scattering amplitude near a resonance.[4] Moreover, the representation (14) is very suitable for averaging procedures over energy intervals of various type, leading in this way to the optical potential for the intermediate structure describing the analogue resonances and to the optical potential for the gross structure.[7]

2) It is suitable to give properties of general character,even if at a formal level, such as

- The TOP is not diagonal in coordinate representation, i.e. it is non-local;
- The expectation value of the imaginary part is negative definite, as it must:

$$\text{(15)} \qquad \mathrm{Im}\langle \mathcal{V}(E)\rangle = -\pi\langle \int d\alpha\, |\langle O\chi|VQ|\Phi_{E,\alpha}\rangle|^2 \quad ,$$

where χ is any state for the relative motion.

- Below the threshold for anelastic processes the real part is monotonically decreasing:

$$\frac{\partial}{\partial E} \mathrm{Re} < \mathcal{V}(E) > < 0 \quad \text{for } E<\varepsilon \tag{16}$$

(nothing can be inferred from Eq. (14) for $E>\varepsilon$, because the integral in Eq. (14) is singular).

- A formal dispersion relation:

$$\mathrm{Re}\,\mathcal{V} = <0|V|0> + \sum_n \frac{<0|VQ|\Phi_n><\Phi_n|QV|0>}{E_n - E} + \frac{1}{\pi} P \int_{\varepsilon}^{+\infty} \frac{\mathrm{Im}\,\mathcal{V}(E')}{E'-E} dE' \tag{17}$$

can be written down. It incorporates the causal and spectral properties.

Of course, all these properties are extablished only at a formal level. If one leaves the formal point of view to assume the mathematical attitude (i.e. all the problems of existence of the various mathematical entities become essential) one discovers that very little has been proved. What has been proved up to now is that, if the nucleon-nucleon potential satisfies certain conditions [3)] which exclude the hard core potentials (such conditions are just those required for the existence of the wave operators in the scattering problem, and may be too severe for a nuclear physicist), the operator $\mathcal{V}(E)$ exists for $\mathrm{Re}\, E > 0$ and it is bounded; moreover its limit for E going to the real axis from the upper complex half plane exists for $E<\varepsilon$. The existence of this limiting value for $E>\varepsilon$, i.e. where the anelastic processes are present, has not been proved up to now and may be it requires more stringent conditions on the nucleon-nucleon potential. Thus the points of view of the mathematics and of the physics, about the TOP, are, at present, very far one from another.

If we now leave the mathematical attitude and return to the formal one, it must be observed that the representation (14) is not useful to calculate the TOP, even in an approximate form, since it requires all the eigenfunctions of the many-body operator QHQ.

2.2.2 The formulation starting from the T-matrix

One can get a possibility of an approximate calculation of the TOP by means of another formulation, which connects the TOP with the nucleon-nucleon scattering amplitude. Such a formulation can be given in various versions: we mention here, in a simplified form, that given by Kerman, Mc Manus and Thaler[8] based on the Lippmann-Schwinger equation for the many-body T-matrix. A similar derivation was given by K.Watson [9] [10] starting from the Lippmann-Schwinger equation for the many-body wave function. The procedure by Kerman, Mc Manus and Thaler consists in eliminating the anelastic channels from the equation:

$$T = V\left(1+(1/d)T\right) \quad , \tag{18}$$

where

$$d = E-H_o-H_N + i\eta \quad , \tag{19}$$

and H_N is the hamiltonian of the target nucleus. From Eq. (18) one obtains two coupled equations for the matrices PTP and QTP; by eliminating QTP one obtains, after some trivial manipulation:

$$PTP = \left(PVP+PVQ\ (d-QVQ + i\eta)^{-1}QVP\right)\left(1+(1/d)PTP\right) \quad , \tag{20}$$

which is a Lippmann-Schwinger equation for the T-matrix projected on the elastic channel subspace, where the operator

$$PVP+PVQ\ (d-QVQ)^{-1}\ QVP \tag{21}$$

plays the role of the potential. Eq. (20) can be put in a two-body form:

$$(22) \qquad \langle 0|T|0\rangle = \left(\langle 0|V|0\rangle + \langle 0|VQ\,(E-QHQ+i\eta)^{-1}\,QV|0\rangle\right) \cdot \left(1 + (1/d_o)\,\langle 0|T|0\rangle\right) \quad ,$$

where

$$(23) \qquad d_o = E - H_o + i\eta$$

(the energy of the ground state of the target has been put equal to zero). Eq. (22) is simply:

$$(24) \qquad \langle 0|T|0\rangle = \mathcal{V}\left(1+(1/d_o)\,\langle 0|T|0\rangle\right) \quad ,$$

where $\mathcal{V}$ is just the theoretical optical potential (11).

The relevance of Eq. (24) lies in the fact that $\mathcal{V}$ can be expanded into a series of nucleon-nucleon scattering amplitude, obtained from the analogous development of the T-matrix, Eq. (18). If V_α is the interaction between the projectile and the α-th target nucleon, and t_α the corresponding scattering operator,

$$(25) \qquad t_\alpha = V_\alpha + V_\alpha\,(1/d)\,t_\alpha \quad ,$$

one obtains, by using

$$(26) \qquad V = \sum_\alpha V_\alpha \; ,$$

from Eq. (18) the development at various orders in the nucleon-nucleon scattering matrix:

$$(27) \qquad T = \sum_\alpha t_\alpha + \sum_{\alpha\neq\beta} t_\alpha (1/d) t_\beta + \sum_{\alpha\neq\beta\neq\gamma} t_\alpha (1/d) t_\beta (1/d) t_\gamma + \ldots . \quad .$$

To the lowest order in t_α, from Eqs. (27) and (24) it follows immediately

$$(28) \qquad \mathcal{V}^{(1)} = \sum_\alpha \langle 0|t_\alpha\,|0\rangle \quad .$$

To get an evaluation of the TOP two important approximations are introduced at this point:

1) The first order approximation (28) for the operator $\mathcal{V}$ ("single scattering approximation").
2) The substitution of the t-matrix for the scattering on a bound nucleon (Eq. (25)) with the t-matrix for the scattering between two free nucleons ("impulse approximation").

Such approximations are expected to be good at high energy.

It must be noted that, if one uses the Kerman-Mc Manus and Thaler formulation, only these two approximations are required to express the TOP in terms of the free nucleon-nucleon scattering amplitude by Eq. (28). Another approximation, which is needed in the derivation by Watson, and whose meaning is rather involved, is not required here.[11)]

The expression (28) is the starting point of a series of manipulations which are straightforward, even if rather heavy, (see, for example, Goldberger and Watson[9)] or Fetter and Watson[10)])which lead to a simpler expression of the TOP in the momentum representation:

(29) $$\langle\vec{k}'|\mathcal{V}|\vec{k}\rangle = \langle\vec{k};\vec{k}'-\vec{k}|\ t(E)\,|\vec{k}\ 0\rangle\ \hat{\rho}(\vec{k}-\vec{k}')\quad ,$$

where:

t(E) is the t-matrix (in the laboratory) for the scattering betwe en two free nucleons at the (laboratory) energy E, averaged over spin and isospin states,

(30) $$\hat{\rho}(\vec{k}-\vec{k}') = \int\exp\left(-i(\vec{k}-\vec{k}').\vec{r}\right)\ \hat{\rho}(\vec{r})d\vec{r}\quad ,$$

and $\hat{\rho}(\vec{r})$ is the nucleon density in the target, normalized to the nucleon number A. With respect to the expression (28), the expression (29) contains the further (minor) approximation of putting equal to zero the initial momentum of the struck nucleon.

Some remarks on the formula (29) may be worthwhile:

1) This formula represents an important step towards an explicit evaluation of the TOP, even if approximate. But the goal to get this one is still very far, because Eq. (29) involves the t-matrix off the energy-shell. To overcome this difficulty some further approximations have been proposed, whose validity, as well as their conceptual meaning, is difficult to be controlled.[12] This point will be considered more in details in two next papers in this meeting.

2) Even in the approximation (29), the TOP in the coordinate representation is non-local. To be local, the off-shell nucleon-nucleon t--matrix would depend on the indices $\vec{k}$ and $\vec{k}'$ only through their difference (momentum transfer). Also this point will be discussed in two next papers later on and it will be shown how Eq. (29) (in fact a rather more accurate expression than this one based on the Galilean invariance [13])can be used to suggest the form of a non-local potential to be employed in the empirical analyses.

2.2.3 Comparison between the two stationary formulations

The two important formulae obtained here are Eq. (14) and Eq. (29). The first is exact, very important to discuss general properties of the target through the spectral representation of the operator QHQ, but it is not suitable for explicit calculations; the second, on the other hand, is a high energy approximation, suitable to suggest reasonable models for the TOP in terms of the off-shell nucleon-nucleon scattering amplitude.

The problem of the antisymmetrization when the projectile is identical with the target nucleons is not a simple one [14] and has been solved in Feshbach's formulation by introducing the correct projection operator P (Ref. 1), which is no longer simply $|0\rangle\langle 0|$, but is given by a rather complicated expression which will not be reproduced here. Starting from the projected many-body Schrödinger equation into the elastic channel subspace, Feshbach derives a wave equation for the asymptotic amplitude for the relative motion where a suitable operator plays the role of the potential.

The main difference with respect to the operator (14) is the presence of a non-local real and energy independent term. It is remarkable that the energy dependence is again of the form shown in Eq. (11), and so a dispersion relation of the type (17) holds again. In conclusion the TOP, when antisymmetrization is taken into account, has the form:

$$(31) \qquad \mathcal{V}\psi(\vec{r}) = \mathcal{V}_1(\vec{r})\psi(\vec{r}) + \int \mathcal{V}_2(\vec{r},\vec{r}\,')\psi(\vec{r}\,')d\vec{r}\,' + \int \mathcal{U}(\vec{r},\vec{r}\,';E)\psi(\vec{r}\,')d\vec{r}\,' \;.$$

Of course the same correct operator P can be used in Eq. (20), obtaining in this way the correct formulation of the TOP in terms of the t-matrix. Starting from this point the corrections to the formula (28) could be evaluated. The argument made in an early paper by Takeda and Watson [8) 15)] which says that Eq. (20) may be used also for identical particles provided that the nucleon-nucleon t-matrix be symmetrized, could be tested.

2.2.4 The simplifications of the previous results in the case of infinite nuclear matter

It is important to see how the expressions (31) and (29) simplify in the limit case of infinite nuclear matter.

Translational and rotational invariance imply that Eq. (31) assumes the form

$$(32) \qquad \mathcal{V}\psi(\vec{r}) = \mathcal{V}_1\psi(\vec{r}) + \int \mathcal{V}_2(|\vec{r}-\vec{r}\,'|)\psi(\vec{r}\,')d\vec{r}\,' + \int \mathcal{U}(|\vec{r}-\vec{r}\,'|,E)\psi(\vec{r}\,')d\vec{r}\,' \;.$$

Moreover the poles appearing in the operator $\mathcal{V}$ (see Eq. (14)) disappear. In the momentum representation Eq. (32) has the form:

$$\hat{\mathcal{V}}(E;\vec{k}) = \hat{\mathcal{V}}_1 \delta(\vec{k}) + \hat{\mathcal{V}}_2(\vec{k}) + \hat{\mathcal{U}}(E,\vec{k}) \qquad (33)$$

where the hat denotes Fourier transform with respect to $\vec{s} = \vec{r}-\vec{r}'$. It must be emphasized that here $\vec{k}$ is the label in the momentum representation, and it is by no means connected with the energy E. The dependence on $\vec{k}$ is the translation, in the momentum representation, of the non-locality in the coordinate representation.

The form assumed by the expression (29) in infinite nuclear matter is:

$$\langle\vec{k}'|\mathcal{V}|\vec{k}\rangle = (2\pi)^3 \rho(0)\delta(\vec{k}-\vec{k}')\langle\vec{k}',\vec{k}-\vec{k}'|t(E)|\vec{k},0\rangle \qquad (34)$$

where the roto-translational invariance is assured by the δ-function, which brings into play only the forward scattering amplitude. It may be worthwhile noting that the δ-function assumed for the nuclear form factor $\hat{\rho}(\vec{k}-\vec{k}')$ does not mean at all that the t-matrix becomes that on the energy-shell: the matrix element $\langle\vec{k}0|t(E)|\vec{k}0\rangle$ is still off energy shell because E and $\vec{k}$ are independent quantities.

Eqs. (33) and (34) (as well as their corresponding for finite nuclei) show that the TOP has a typical energy dependence: in Eq. (33) it is contained in the resolvent of the operator QHQ; in Eq. (34) it is given by the energy dependence of the t-matrix for nucleon-nucleon scattering. Such an energy dependence is also referred to as the "dynamical" or "intrinsic",[16] in order to distinguish it from an additional energy dependence which comes into play when one trasforms the TOP into a local phase-equivalent potential. Such a point will be discussed in the next paper in this meeting.

3. The formulation of the theoretical optical potential by means of the one-particle Green function

3.1 The connection between the Green function and the theoretical optical potential

I shall now sketch briefly how techniques of the many-body theory of nuclear matter can be used to calculate the TOP. This point will be treated in more detail by prof. Mahaux and so I shall only mention here the essential connections.

I shall follow the pioneering work by Bell and Squires.[17)]

Consider a state of the many-body system at the time t in the coordinate representation

(35) $$|\Psi_-\rangle = \int \phi_-(\vec{r}) \quad \psi^+(\vec{r}\,)|0\rangle \, d\vec{r} \quad ,$$

where, as above, $|0\rangle$ is the vector describing the correlated ground state of the target, $\psi^+(\vec{r})$ is the Schrödinger creation operator at the position $\vec{r}$, and therefore $\phi_-(\vec{r})$ is the wave function for the relative motion at the time t. Let us consider another state at the time $t'>t$:

(36) $$|\Psi_+\rangle \;=\; \int \phi_+(r)\psi^+(\vec{r})|0\rangle \, d\vec{r} \quad ,$$

and express the probability amplitude for finding, at the time t', the state $|\Psi_+\rangle$ into the state evolved from $|\Psi_-\rangle$:

(37) $$\langle\Psi_+|e^{-iH(t'-t)}|\Psi_-\rangle = \int \phi_+^*(\vec{r}\,')\phi_-(\vec{r})\langle 0|\psi(\vec{r}\,')e^{-iH(t'-t)} \psi^+(\vec{r})|0\rangle \, d\vec{r} \, d\vec{r}\,' \quad (t'>t) \; .$$

Of course such an amplitude vanishes for $t'<t$: such a condition is typical of a scattering process and is a trivial form of the causality condition. If $\phi_-(\vec{r})$ and $\phi_+(\vec{r})$ describe wave packets very far from the

nucleus, Eq. (37) describes what we shall call an elastic scattering amplitude for finite times. We consider finite times in order not to complicate the following discussion with the subtilities implied by the limits for these times going to $\pm\infty$, which are unessential to our present purpose. Thus the quantity given by Eq. (37) is not an S-matrix element.

Recall now the expression of the one-particle Green function:

$$G(\vec{r}',\vec{r};t'-t) = -i\langle 0|T(\psi(\vec{r}',t')\psi^+(\vec{r},t))|0\rangle \quad , \tag{38}$$

where T is the usual time-ordered product, ψ and ψ^+ are Heisenberg operators. The Green function can be splitted in the so called particle (G_p) and hole (G_h) parts:

$$G = G_p + G_h \quad , \tag{39}$$

where:

$$\begin{aligned} G_p(\vec{r}',\vec{r};t'-t) &= -i\langle 0|\psi(\vec{r}',t')\psi^+(r,t)|0\rangle \quad , \quad t'-t>0 \\ &= 0 \quad , \quad t'-t<0 \quad , \end{aligned} \tag{40}$$

$$\begin{aligned} G_h(\vec{r}',\vec{r},t'-t) &= 0 \quad , \quad t'-t>0 \\ &= i\langle 0|\psi^+(\vec{r},t)\psi(\vec{r}',t')|0\rangle \quad , \quad t'-t<0 \ . \end{aligned} \tag{41}$$

After transforming the operators from the Heisenberg to the Schrödinger picture, it is easily recognized that Eq. (37) can be written as (the energy of the ground state of the target has been put equal to zero):

$$\langle\Psi_+|e^{-iH(t'-t)}|\Psi_-\rangle = \int\phi_+^*(\vec{r}')\phi_-(\vec{r})G_p(\vec{r}',\vec{r},t'-t)d\vec{r}\,d\vec{r}' \quad , \tag{42}$$

and so the function $G_p(\vec{r}',\vec{r};t'-t)$ has the meaning of the Green function for the two body problem of the relative motion.

At this point a very important observation by Bell and Squires must be recalled: as long as the wave packets $\phi_+(\vec{r})$ and $\phi_-(\vec{r})$ do not overlap with target wave function, nothing is changed by substituting, in Eq. (42), the function G_p by the function G.

Now the question is how this meaning of the whole one-particle Green function can be used to calculate the TOP. The whole one-particle Green function satisfies the Dyson equation:

$$G = G_o + G_o M G \ , \tag{43}$$

where G_o is the one-particle Green function for the uncorrelated system and M the proper self energy (or mass operator). Such an equation is of Lippmann-Schwinger type and the operator M plays the role of the potential. For this reason the mass operator M may be considered as a potential giving the correct nucleon-nucleus scattering amplitudes. Such a conclusion is very important, both for computational and for conceptual reasons:

1) The operator M can be calculated, for nucleon-nucleon realistic interactions, by techniques developed in the nuclear many-body theory, as, for instance, the hole-line expansion. In this way properties such as the energy dependence of its real and imaginary parts and its non-locality, which are quite involved in the representations (14) and (29), may be explicitely evaluated.[18,19] These results will be shown in a next paper by prof. Mahaux.

2) It opens the more straightforward way to extend the concept of the theoretical optical potential, here introduced for scattering processes, also to negative energies, describing in this way the simple particle excitations in a nucleus.

A direct connection between the mass operator M and the theoretical optical potential introduced in Sect. 2 can be easily seen in the high energy limit.[18] In the lowest order in the hole-line expansion (Brueckner-Hartree-Fock approximation) the mass operator is given by, in the momentum representation:

$$M(k;E) = \sum_{j<k_F} \langle \vec{k}\vec{j} | g(E+e(\vec{j})) | \vec{k}\vec{j} \rangle , \tag{44}$$

where $e(\vec{j})$ is the energy of a particle of momentum $\vec{j}$ in nuclear matter, g is the reaction matrix which satisfies the Bethe-Goldstone equation. In the high energy limit

---

Bethe-Goldstone equation $\rightarrow$ Lippmann-Schwinger equation

---

Reaction g-matrix $\langle \vec{k}\vec{j} | g(E+e(\vec{j})) | \vec{k}\vec{j} \rangle \rightarrow$ scattering t-matrix $\langle \vec{k}0 | t(E) | \vec{k}0 \rangle$

$$\sum_{j<k_F} \longrightarrow \text{factor } \frac{4\pi}{3} k_F^{\,3} = (2\pi)^3 \rho$$

---

and thus from Eq. (44) the high energy approximation (29) is obtained.

References

1) H.Feshbach, Ann. Phys. 19,287 (1962).
2) F.Coester and H.Kümmel, Nucl. Phys. 9,225 (1958).
3) M.Bertero and G.Passatore, Nuovo Cimento 2A,579 (1971).
4) H.Feshbach, Ann. Phys. 5,357 (1958).
5) A.M.Lane and R.G.Thomas, Revs. Modern Phys. 30,257 (1958).
6) G.Brown, Rev. Mod. Phys. 31,893 (1959).
7) H.Feshbach, A.K.Kerman and R.H.Lemmer, Ann. Phys. 41, 230 (1967).
8) A.K.Kerman, H. Mc Manus and R.M.Thaler, Ann.Phys. 8,551 (1959).
9) M.L.Goldberger and K.Watson, Collision Theory, Wiley and Sons,

N.Y. (1964).

10) A.Fetter, K.Watson, Adv. Theor. Phys. 1,115 (1965).

11) M.Giannini, Nuovo Cimento 3A,365 (1971).

12) R.E.Schenter and B.W. Downs, Phys. Rev. 133B,522 (1964).

13) M.Giannini and G.Ricco, to be published.

14) J.S.Bell, Lectures on many body problem, (1959).

15) G.Takeda and K.Watson, Phys. Rev. 97,1336 (1955).

16) R.Lipperheide, Nucl. Phys. 89,97 (1966).

17) J.S.Bell and E.J.Squires, Phys. Rev.Lett. 3,96 (1959).

18) J.P.Jeukenne, A.Lejeune and C.Mahaux, to be published.

19) J.R.Rook, Nucl. Phys. A222,596 (1974).

EQUIVALENT POTENTIALS IN THE DESCRIPTION OF SCATTERING PROCESSES

F. CAPUZZI

Istituto di Fisica Teorica, Università di Pavia, Pavia
Istituto Nazionale di Fisica Nucleare, Sezione di Pavia

Abstract. Some connections among nonlocal, almost local and local descriptions of scattering processes are investigated. An extension of the Perey-Saxon expansion for the Frahn-Lemmer model gives formally an equivalent potential whose nonlocality is restricted to the nuclear surface. Hence, a local phase-equivalent potential and a corresponding almost local Perey effect are found to the second order of the surface terms. Under general conditions on the nonlocal optical potential an almost local equivalent potential of the Kisslinger's form is obtained. This is parity-conserving and reproduces exactly the nonlocal wave functions. Some general properties (like breakdown of causality and of spherical symmetry) are investigated from a qualitative point of view.

1. Introduction

Some different theoretical approaches to the particle-nucleus elastic scattering[1-4] introduce in the Schrödinger equation a nonlocal potential $\mathcal{V}(E;\underline{r},\underline{r}')$ (theoretical optical potential):

(1.1) $$(\underline{\nabla}^2+E)\Psi(\underline{q},\underline{r}) = \int \mathcal{V}(E;\underline{r},\underline{r}')\Psi(\underline{q},\underline{r}')d\underline{r}', \quad \hbar = 2\mu=1,\ q^2 = E.$$

Here, $\Psi(\underline{q},\underline{r})$ denotes the scattering solution, which behaves at infinity as follows:

(1.2) $$\Psi(\underline{q},\underline{r}) \overset{r\to\infty}{\simeq} \exp(i\underline{q}.\underline{r})+f(E,\Theta)\ \frac{\exp(iqr)}{r}, \Theta = \text{scattering angle}.$$

The integral kernel $\mathcal{V}(E;\underline{r},\underline{r}')$ (that is complex, in general) is a symmetrical function of $\underline{r}$ and $\underline{r}'$, as a consequence of the time-reversal inva riance.[5] The causal origin of its energy dependence (named usually dynamical energy dependence) is the source of a dispersion relation that connects the real and the imaginary part of $\mathcal{V}(E;\underline{r},\underline{r}')$ (Ref.3).

Despite of the nonlocality of the theoretical optical potential, the experimental data are analysed usually by means of a local complex potential (phenomenological optical potential). This shows both a dynamical and a spurious energy dependence; hence, it does not satisfy a dispersion relation.[6,7] We use the following symbols in the corresponding Schrödinger equation:

$$(1.3) \qquad (\underline{\nabla}^2+E)\Psi_L(\underline{q},\underline{r}) = \mathcal{V}_L(E,r)\Psi_L(\underline{q},\underline{r}) \quad .$$

Besides, in some phenomenological analyses, even an "almost local" form of potential is used. This is the so-called Kisslinger potential, which yields the Schrödinger equation:

$$(1.4) \qquad (\underline{\nabla}^2+E)\phi(\underline{q},\underline{r}) = \{A(E,r) - \underline{\nabla}\cdot(B(E,r)\underline{\nabla})\}\ \phi(\underline{q},\underline{r}) \quad .$$

This potential was deduced first by Kisslinger [8] as an approximation to the theoretical optical potential in multiple scattering theory. It has given successful fits to low energy experimental data in pion-nucleus elastic scattering and it has found an increasing theoretical and phenomenological interest in the past few years.[9-13]

The wide use of local and almost local potentials in the phenomenological analyses makes worthwhile an investigation of their connections with the "true" nonlocal interaction. In this note we are concerned with the problem of constructing a local or an almost local potential that is equivalent to a nonlocal one, i.e. that reproduces the same observable features. Since our interest here is devoted mainly to the scattering problem, "equivalent to" means in all the paper "that gives the same scattering amplitudes as". To be more precise, we make a distinction between a full and a phase equivalence.

A potential $\mathcal{E}(E;\underline{r},\underline{r}')$ is said to be full equivalent to $\mathcal{V}(E;\underline{r},\underline{r}')$ if for every scattering solution $\Psi(\underline{q},\underline{r})$ of $\mathcal{V}(E;\underline{r},\underline{r}')$ it is

$$\int \mathcal{E}(E;\underline{r},\underline{r}')\Psi(\underline{q},\underline{r}')d\underline{r}' = \int \mathcal{V}(E;\underline{r},\underline{r}')\Psi(\underline{q},\underline{r}')d\underline{r}' \tag{1.5}$$

on the energy shell $q^2 = E$. Therefore, $\mathcal{E}(E;\underline{r},\underline{r}')$has the same scattering solutions as $\mathcal{V}(E;\underline{r},\underline{r}')$, but its analytical form and, hence, its action on other wave functions is different.

On the contrary, a potential is only phase-equivalent to $\mathcal{V}(E;\underline{r},\underline{r}')$ if it gives origin to scattering solutions $\chi(\underline{q},\underline{r})$ which are different from $\Psi(\underline{q},\underline{r})$ but which have the same asymptotic behaviour:

$$\chi(\underline{q},\underline{r}) \overset{r\to\infty}{\simeq} \exp(i\underline{q}.\underline{r})+f(E,\Theta)\ \frac{\exp(iqr)}{r} \ . \tag{1.6}$$

We give two examples of the above definitions.

If both the sides of Eq. (1.4) are divided by $1+B(E,r)$, one obtains

$$(\underline{\nabla}^2+E)\phi(\underline{q},\underline{r}) = \left[\tilde{A}(E,r)+(\underline{\nabla}\ \tilde{B}(E,r)).\underline{\nabla}\right]\phi(\underline{q},\underline{r}) \quad , \tag{1.7}$$

where

$$\tilde{A}(E,r) = \frac{A(E,r)+B(E,r)E}{1+B(E,r)} \quad , \quad \tilde{B}(E,r) = -\ln\left[1+B(E,r)\right] \ . \tag{1.8}$$

The new potential provides the same solutions as the first, but its integral kernel is not symmetrical and it shows a different energy dependence. In the following it will be denominated "not symmetrical Kisslinger potential".

A simple example of phase equivalence between the Kisslinger potential and a local one is due to Fiedeldey.[14] Let us set

$$\phi(\underline{q},\underline{r}) = \left[1+B(E,r)\right]^{-1/2}\Psi_L(\underline{q},\underline{r}) \tag{1.9}$$

and insert it in Eq. (1.4). Therefore one gets

$$(\underline{\nabla}^2+E)\Psi_L(\underline{q},\underline{r}) = \mathcal{V}_L(E,r)\Psi_L(\underline{q},\underline{r}) \ , \tag{1.10}$$

where

$$(1.11) \qquad \mathcal{V}_L(E,r) = \frac{A(E,r)+B(E,r)E + \frac{1}{2}\nabla^2 B(E,r)}{1+B(E,r)} - \frac{1}{4}\left(\frac{\nabla B(E,r)}{1+B(E,r)}\right)^2 .$$

Now, $\phi(\underline{q},\underline{r})$ and $\Psi_L(\underline{q},\underline{r})$ are obviously phase-equivalent, provided that it is $\lim_{r\to\infty} rB(E,r)=0$. The relation between $\phi(\underline{q},\underline{r})$ and $\Psi_L(\underline{q},\underline{r})$ is called Perey effect. [15)]

Before considering a possible equivalence among the above potentials, a preliminary remark is needed. In fact, if one compares the corresponding scattering theories,* the existence of spurious states appears to be a very delicate point.

There are bound states with a positive energy which may arise in the low energy region when one solves a nonlocal Schrödinger equation. [19)] On the contrary, no such a state exists for a local or an almost local potential, provided that this is a regular one. Such a different behaviour influences also the scattering problem, since the boundary condition (1.2) does not provide a unique solution, if this corresponds to a spurious state. In this case a local (as well as an almost local) equivalent potential either loses some solutions or it is not expected to have a mathematical convenience (since it is singular or multivalued).

In Sect. 2 we deduce from the Frahn-Lemmer model a full equivalent potential which is nonlocal only at the nuclear surface. The method that we adopt is a clarification and an extension of the Perey-Saxon expansion. Then, we examine the case of a gaussian nonlocal form factor (Sect. 3); from the equivalent potential derived in Sect. 2 we obtain to the second order in the surface effects a local phase-equivalent po

---

* A rigorous scattering theory for local and nonlocal potentials is found in Refs. 16 to 20; the almost local case is directly related to the local one by means of Eq. (1.9).

tential and the corresponding Perey effect.

In Sect. 4, under general conditions on the theoretical optical potential, we derive in an exact way a full equivalent potential of the Kisslinger's form. Some properties of such a potential (like breakdown of causality and of spherical symmetry) are investigated from a qualitative point of view.

2. A full equivalent potential for the Frahn-Lemmer model

In this Section we consider the simplified [*] Frahn-Lemmer model for the theoretical optical potential [21-23] :

$$(2.1) \qquad \mathcal{V}_{FL}(E;\underline{r},\underline{r}') = H(s)U(E,S) \;, \quad \underline{s} \equiv \underline{r}-\underline{r}' \;, \; \underline{S} = \frac{\underline{r}+\underline{r}'}{2} \;.$$

Here, H(s) is a nonlocal form factor whose range can be taken as a measure of the nonlocality; U(E,S) is supposed to have a Saxon-Woods form. In order to simplify our notations, the dynamical energy dependence of U(E,S) will be omitted in the following.

With a view to expressing the potential in a more convenient form, we introduce the Fourier transforms

$$(2.2) \qquad F\left(-\left(\frac{\underline{k}+\underline{k}'}{2}\right)^2\right) \equiv \hat{H}\left(\frac{\underline{k}+\underline{k}'}{2}\right) = \int \exp\left(-i\left(\frac{\underline{k}+\underline{k}'}{2}\right).\underline{s}\right) H(s)\, d\underline{s} \;,$$

$$(2.3) \qquad \hat{U}(\underline{k}-\underline{k}') = (2\pi)^{-3/2} \int \exp\left(-i(\underline{k}-\underline{k}').\underline{S}\right) U(S)\, d\underline{S} \;,$$

$$(2.4) \qquad \hat{\Psi}(\underline{q},\underline{k}') = (2\pi)^{-3/2} \int \exp(-i\underline{k}'.\underline{r}')\, \Psi(\underline{q},\underline{r}')\, d\underline{r}' \;,$$

---

[*] We do not consider the spin-orbit contribution.

and then we write the Schrödinger equation for $\mathcal{V}_{FL}(\underline{r},\underline{r}')$ in the following way:

$$(2.5) \qquad (\underline{\nabla}^2+E)\Psi(\underline{q},\underline{r}) = \int \mathcal{V}_{FL}(\underline{r},\underline{r}')\Psi(\underline{q},\underline{r}')d\underline{r}'$$

$$= (2\pi)^{-3}\int \exp(i\underline{k}.\underline{r})F\left(-\left(\frac{\underline{k}+\underline{k}'}{2}\right)^2\right)\hat{U}(\underline{k}-\underline{k}')\hat{\Psi}(\underline{q},\underline{k}')d\underline{k}\; d\underline{k}' .$$

By means of the well known correspondences

$$(2.6) \qquad \begin{aligned} \underline{k}-\underline{k}' &\sim -i\underline{\nabla}' , \quad \text{where} \quad \underline{\nabla}' \text{ acts on } \underline{r} \text{ in } U(r) , \\ \underline{k}' &\sim -i\underline{\nabla}'' , \quad \text{where} \quad \underline{\nabla}'' \text{ acts on } \underline{r} \text{ in } \Psi(\underline{q},\underline{r}) , \end{aligned}$$

we get formally

$$(2.7) \qquad \int \mathcal{V}_{FL}(\underline{r},\underline{r}')\Psi(\underline{q},\underline{r}')d\underline{r}' = F\left(\left(\frac{\underline{\nabla}'}{2} + \underline{\nabla}''\right)^2\right)U(r)\Psi(\underline{q},\underline{r}) .$$

Thus, the action of $\mathcal{V}_{FL}$ on Ψ can be described as the action of a function F of two commuting operators on $U\Psi$ [x].

Perey and Saxon [24] expand F about a local operator and then they stop the development to the first order. Starting from this idea, we elaborate a more well defined procedure in order to clarify the problem and to achieve higher approximations. First we observe that, in order to take into account better the nonlocality of $\mathcal{V}_{FL}$, one must expand F about a nonlocal operator Λ which can be assumed to act directly on Ψ. Then, we use the Taylor development

$$(2.8) \qquad F(O) = F(\Lambda)+(O-\Lambda)F'(\Lambda)+ \frac{1}{2}(O-\Lambda)^2F''(\Lambda)+ \frac{1}{2}[\Lambda,O]F''(\Lambda)+\ldots ,$$

where O denotes the differential operator $(\underline{\nabla}'/2 + \underline{\nabla}'')^2$.

---

x Here and in the following we use compact notations for the operators and the wave functions.

If we arrange in a suitable way the terms of the above expansion, the following expression for $\mathcal{V}_{FL}$ is obtained

(2.9) $$\mathcal{V}_{FL} = \mathcal{F}(\Lambda) + \mathcal{G}(\Lambda) \quad ,$$

where

(2.10) $$\mathcal{F}(\Lambda) = UF(\Lambda) - \frac{1}{4}(\underline{\nabla}^2 U)F'(\Lambda) + \frac{1}{32}(\underline{\nabla}^4 U)F''(\Lambda)$$
$$+ \frac{1}{2}\,[\underline{\nabla}^2, U]\,F'(\Lambda) - \frac{1}{8}[\underline{\nabla}^2,(\underline{\nabla}^2 U)]\,F''(\Lambda) + \frac{1}{2}U[\underline{\nabla}^2, \Lambda]F''(\Lambda)$$
$$+ \frac{1}{8}\,[\underline{\nabla}^2,[\underline{\nabla}^2, U]]\,F''(\Lambda) + \ldots \quad ,$$

(2.11) $$\mathcal{G}(\Lambda) = UF'(\Lambda)\,(\underline{\nabla}^2-\Lambda) + \frac{1}{2}UF''(\Lambda)\,(\underline{\nabla}^2-\Lambda)^2 + \frac{1}{2}[\underline{\nabla}^2, U]F''(\Lambda)(\underline{\nabla}^2-\Lambda)$$
$$+ \ldots\ldots\ldots\ldots\ldots\ldots \quad .$$

This partition separates a term $\mathcal{F}(\Lambda)$, where the Laplacian appears only inside the commutators, from a term $\mathcal{G}(\Lambda)$, where the operator $\underline{\nabla}^2-\Lambda$ acts directly on the wave function. Our purpose consists in eliminating $\mathcal{G}(\Lambda)$ by means of a suitable choice of Λ , that is drawn from the following considerations.

Let ϕ be a solution of the Schrödinger equation

(2.12) $$(\underline{\nabla}^2+E)\,\phi = \mathcal{E}(E)\,\phi \quad ,$$

where we have set

(2.13) $$\mathcal{E}(E) = \mathcal{F}(\Lambda(E)) \quad ,$$

(2.14) $$\Lambda(E) = \mathcal{E}(E) - E \quad .$$

Therefore, by means of Eqs. (2.9) and (2.11)-(2.14), one gets

(2.15) $$(\mathcal{V}_{FL} - \mathcal{E}(E))\,\phi = \mathcal{G}(\Lambda(E))\,\phi = 0$$

and then

(2.16) $$(\underline{\nabla}^2+E)\,\phi = \mathcal{V}_{FL}\,\phi \;.$$

Thus, $\mathcal{E}(E)$ is full equivalent to $\mathcal{V}_{FL}$, since every solution of Eq. (2.12) is also a solution of Eq. (2.16). Nevertheless, it must be emphasized that some solutions of Eq. (2.16) may be missing in the equivalent one. Therefore, the equivalent equation requires some caution, since one must make sure that the scattering solution exists.

Eqs. (2.13) and (2.14) give the following self-consistent definitions of $\mathcal{E}(E)$ and $\Lambda(E)$:

$$(2.17) \qquad \mathcal{E}(E) = \mathcal{Y}(\mathcal{E}(E)-E) \quad ,$$

$$(2.18) \qquad \Lambda(E) = \mathcal{Y}(\Lambda(E)) - E \quad .$$

The choice of $\mathcal{E}(E)$ instead of $\mathcal{V}_{FL}$ offers the practical advantage that the main part of the nonlocality has been eliminated. In fact, $\mathcal{G}(\Lambda(E))$ acts in a nonlocal way on the whole nuclear region. On the contrary , the nonlocality of $\mathcal{E}(E)$ arises only from the commutators and from its dependence upon $\Lambda(E)$. This last nonlocality, as we will see later solving Eq. (2.17), is connected with similar commutators. Now, these are different from zero only at the nuclear surface, because it is

$$(2.19) \qquad \underbrace{[\underline{\nabla}^2, [\underline{\nabla}^2, \ldots . [\underline{\nabla}^2, U]]]}_{n}\Psi = \left(\underline{\nabla}' . (\underline{\nabla}' + 2\underline{\nabla}'')\right)^n U\Psi \quad .$$

Therefore on the physical ground we can expect a fast convergence of the development. [x]

Of course, we pay a price for this advantage. In fact, $\mathcal{E}(E)$ has not a symmetrical kernel and, moreover, it shows a spurious energy dependen-

---

[x] We remark that a different full equivalent potential $\tilde{\mathcal{E}}(E)$ can be obtained from the choice $\tilde{\Lambda}(E) = \mathcal{V}_{FL}-E$, that gives the explicit equation $\tilde{\mathcal{E}}(E) = \mathcal{Y}(\mathcal{V}_{FL}-E)$. This potential is not useful to our purpose because its nonlocality is not restricted to the nuclear surface. We observe also that $\tilde{\mathcal{E}}(E)$ can give more solutions than $\mathcal{V}_{FL}$.

ce (indicated explicitely in our symbols) that originates from its dependence upon $\Lambda(E)$.

In the above treatment, no attempt at mathematical rigour has been made. Thus, a list of open problems follows. Firstly, the convergence of the expansion (2.10) is not proved. Secondly, if this expansion converges, $\mathcal{E}(E)$ has not a symmetrical kernel and nothing is known about the scattering solutions for such a potential. This is a delicate point that is connected directly with the problem of an eventual loss of solutions when $\mathcal{V}_{FL}$ is replaced by $\mathcal{E}(E)$. Thirdly, no condition is at our disposal in order to justify well a truncation of the development. Moreover, if one considers commutators of order higher than the second, even the order of the Schrödinger equation increases and this fact disguises nontrivial mathematical problems.

In practice, the expansion (2.10) can be considered only up to the second order. In this case, the existence of the scattering solution follows from the construction of a local phase-equivalent potential, as we shall see in the next Section.

3. A local phase-equivalent potential for the Frahn-Lemmer model (gaussian case)

In this Section we confine ourselves to the gaussian nonlocal form factor, which is used in some phenomenological analyses [22,23]:

$$(3.1) \qquad H(s) = \pi^{-3/2}a^{-3}\exp\left(-\left(\frac{s}{a}\right)^2\right), \; F(k^2) = \exp\left(\alpha k^2\right) \; , \; \alpha = \frac{a^2}{4} \; ,$$

where a is the range of the nonlocality. This restriction is not essential to our purpose, but it is useful in order to obtain more compact formulas.

We deal with the equivalent Schrödinger equation

$$(2.12) \qquad (\underline{\nabla}^2 + E)\phi = \mathcal{E}(E)\phi .$$

In the gaussian case the equivalent potential $\mathcal{E}(E)$, by means of Eqs. (2.17) and (2.10), satisfies the implicit equation

$$(3.2) \qquad \mathcal{E}(E) = \Big(\tilde{U} + \frac{\alpha}{2}[\underline{\nabla}^2, \tilde{U}] + \frac{\alpha^2}{2}\tilde{U}[\underline{\nabla}^2, \mathcal{E}(E)] + \frac{\alpha^2}{8}[\underline{\nabla}^2, [\underline{\nabla}^2, \tilde{U}]] + \ldots\Big) \exp\big(\alpha(\mathcal{E}(E) - E)\big) ,$$

where, collecting some terms of the expansion (2.10), we have set

$$(3.3) \qquad \tilde{U} = \exp\Big(-\frac{\alpha}{4}\underline{\nabla}^2\Big)U .$$

Now we truncate the above development to the commutators of the second order and then we solve the equation to a corresponding approximation. For this purpose we observe that, if all the commutators are neglected, one obtains a local solution $\tilde{V}(E)$ which satisfies the implicit equation

$$(3.4) \qquad \tilde{V}(E) = \tilde{U}\exp\big(\alpha(\tilde{V}(E) - E)\big) .$$

This suggests a solution of the form

$$(3.5) \qquad \mathcal{E}(E) = \tilde{V}(E) + \alpha\mathcal{E}_1(E) ,$$

where $\mathcal{E}_1(E)$ is a nonlocal surface term which can be obtained by substituting Eq. (3.5) into Eq. (3.2). If $\mathcal{E}_1(E)$ is taken into account up to the second order of α, the following expression for the equivalent potential is obtained:

$$(3.6) \qquad \mathcal{E}^{(2)}(E) = \tilde{V}(E) + \big(1 - \alpha\tilde{V}(E)\big)^{-1}\Big(\frac{\alpha}{2}[\underline{\nabla}^2, \tilde{U}] + \frac{\alpha^2}{2}\tilde{U}[\underline{\nabla}^2, \tilde{V}(E)] + \frac{\alpha^2}{8}[\underline{\nabla}^2, [\underline{\nabla}^2, \tilde{U}]]\Big) \exp\big(\alpha(\tilde{V}(E) - E)\big) .$$

As concerns the energy dependence of $\mathcal{E}^{(2)}(E)$, we observe that it arises from $\tilde{U}$ (dynamical dependence that we omit in our symbolism) and from $\tilde{V}(E)$. This last depends upon the energy both in a dynamical and in a spurious way. The spurious energy dependence is indicated explicitely and originates from the factor $\exp(-\alpha E)$ in Eq. (3.4).

Before constructing a local potential phase-equivalent to $\mathcal{E}^{(2)}(E)$, we examine shortly the lower approximations in order to recover the results of Perey-Buck and Fiedeldey.

a) Zeroth order or Perey-Buck approximation

Let us neglect in Eqs. (3.3), (3.4) and (3.6) all the surface contributions (as it is the case when infinite nuclear matter is considered). Therefore, we obtain

$$\mathcal{E}^{(o)}(E) = V(E) \quad , \tag{3.7}$$

where $V(E)$ is the local solution of the Perey-Buck self-consistent equation [23]

$$V(E) = U \exp\left[\alpha(V(E)-E)\right] . \tag{3.8}$$

b) First order approximation

If the surface terms of order higher than the first are neglected in Eqs. (3.3), (3.4) and (3.6), one obtains

$$\tilde{V}(E) \simeq V(E) - \frac{\alpha}{4} \frac{\underline{\nabla}^2 U}{1-\alpha V(E)} \exp\left[\alpha(V(E)-E)\right] \tag{3.9}$$

and then

$$\mathcal{E}^{(1)}(E)=V(E)+ \alpha\frac{\exp\left[\alpha(V(E)-E)\right]}{1-\alpha V(E)}\left(\frac{1}{4}(\underline{\nabla}^2 U)+(\underline{\nabla}U)\cdot\underline{\nabla}\right)=V(E)+\frac{\alpha}{4}(\underline{\nabla}^2 V(E)) - \frac{\alpha^2}{4}\frac{2-\alpha V(E)}{1-\alpha V(E)}(\underline{\nabla}V(E))^2 + \alpha(\underline{\nabla}V(E)).\underline{\nabla} , \tag{3.10}$$

where $\mathcal{E}^{(1)}(E)$ has been expressed in terms of the Perey-Buck potential

V(E) by means of Eq. (3.8).

In this approximation the equivalent potential $\mathcal{E}(E)$ is a nonsymmetrical Kisslinger potential. Its symmetrical equivalent can be obtained by inverting Eqs. (1.8). Thus, we write the following Schrödinger equation

$$(\underline{\nabla}^2+E)\phi = \left(A(E) - \underline{\nabla}\cdot(B(E)\underline{\nabla})\right)\phi \quad , \tag{3.11}$$

where

$$A(E)=\left(V(E)+\frac{\alpha}{4}(\underline{\nabla}^2 V(E))-\frac{\alpha^2}{4}\,\frac{2-\alpha V(E)}{1-\alpha V(E)}\,(\underline{\nabla}V(E))^2\right)\exp\left(-\alpha V(E)\right)$$
$$+\left(1-\exp\left(-\alpha V(E)\right)\right)E \quad , \tag{3.12}$$

$$B(E) = \exp\left(-\alpha V(E)\right)-1 \quad . \tag{3.13}$$

A local potential phase-equivalent to $\mathcal{E}^{(1)}(E)$ and the corresponding Perey effect are obtained directly from Eqs. (1.11) and (1.9) that give the formulas of Fiedeldey:[14)]

$$\mathcal{V}_L(E) = V(E)-\frac{\alpha}{4}(\underline{\nabla}^2 V(E))-\frac{\alpha^2}{4}\left(1-\alpha V(E)\right)^{-1}(\underline{\nabla}V(E))^2 \quad , \tag{3.14}$$

$$\phi = \exp\left(\frac{\alpha}{2}\,V(E)\right)\Psi_L \quad . \tag{3.15}$$

c) Second order approximation

If we neglect some surface corrections higher than α^2, Eq. (3.6) can be written in a more compact way in terms of $\tilde{V}(E)$:

$$\mathcal{E}^{(2)}(E)\simeq \tilde{V}(E)+\left(1-\alpha\tilde{V}(E)\right)^{-1}\left(\frac{\alpha}{2}\,[\underline{\nabla}^2,\tilde{V}(E)]+\frac{\alpha^2}{8}[\underline{\nabla}^2,[\underline{\nabla}^2,\tilde{V}(E)]]\right). \tag{3.16}$$

In order to obtain a phase-equivalent local potential we make use of the following argument.

Let us consider the Schrödinger equation

$$(\underline{\nabla}^2+E)\phi = (V_o + \alpha[\underline{\nabla}^2, V_1] + \alpha^2[\underline{\nabla}^2,[\underline{\nabla}^2, V_2]])\phi \quad , \tag{3.17}$$

where V_o, V_1 and V_2 are local operators and it is

$$\lim_{r\to\infty} rV_1 = \lim_{r\to\infty} r \frac{\partial V_2}{\partial r} = 0 \quad . \tag{3.18}$$

For every solution Ψ_L of the local equation

$$(\underline{\nabla}^2 + E)\, \Psi_L = \mathcal{V}_L \Psi_L \quad , \tag{3.19}$$

where

$$\mathcal{V}_L = V_o + \alpha^2(\underline{\nabla}V_1)^2 - 2\alpha^2(\underline{\nabla}V_o)\cdot(\underline{\nabla}V_2) \quad , \tag{3.20}$$

the function

$$\phi = \exp(\alpha V_1)(1+\alpha^2[\underline{\nabla}^2, V_2])\, \Psi_L \tag{3.21}$$

satisfies Eq. (3.17) to the second order in α. As a consequence of the conditions (3.18), ϕ is phase-equivalent to Ψ_L.

A direct substitution of ϕ into Eq. (3.17) and some easy calculations of commutators supply the proof of this statement.

Now we write $\mathcal{E}^{(2)}(E)$ as a sum of commutators. Setting

$$V_o(E) = \tilde{V}(E) - 2(\underline{\nabla}\ln(1-\alpha\tilde{V}(E))^{1/2})^2, \tag{3.22}$$

$$V_1(E) = -\frac{1}{\alpha}\ln(1-\alpha\tilde{V}(E))^{1/2} \quad , \tag{3.23}$$

$$V_2(E) = -\frac{1}{4\alpha}\ln(1-\alpha\tilde{V}(E))^{1/2} \quad , \tag{3.24}$$

$\mathcal{E}^{(2)}(E)$ becomes, to the second order of the surface terms,

$$\mathcal{E}^{(2)}(E) = V_o(E) + \alpha[\underline{\nabla}^2, V_1(E)] + \alpha^2[\underline{\nabla}^2,[\underline{\nabla}^2, V_2(E)]] \quad . \tag{3.25}$$

Thus, from Eqs. (3.20) and (3.21), we obtain the following expressions for the equivalent local potential and the corresponding Perey effect:

$$(3.26)\qquad \mathcal{V}_L(E) = \tilde{V}(E) - \frac{\alpha^2}{2}\left(\frac{\underline{\nabla}\tilde{V}(E)}{1-\alpha\tilde{V}(E)}\right)^2 ,$$

$$(3.27)\qquad \phi = \left(1-\alpha\tilde{V}(E)\right)^{-1/2}\left(1+\frac{\alpha^2}{8}\frac{\underline{\nabla}^2\tilde{V}(E)}{1-\alpha\tilde{V}(E)} + \frac{\alpha^2}{4}\frac{\underline{\nabla}\tilde{V}(E)}{1-\alpha\tilde{V}(E)}\cdot\underline{\nabla}\right)\Psi_L ,$$

where similar surface contributions of the third order have been dropped.

Apart from a second order surface correction, the equivalent local potential coincides with the solution $\tilde{V}(E)$ of Eq. (3.4). This points out that the Perey-Buck self-consistent relation is significant also for finite nuclei, provided that the potential function U is replaced by $\tilde{U}$ (see Eqs. (3.3), (3.4) and (3.8)).

In order to obtain a direct comparison with the results of Fiedeldey, we express $\tilde{V}(E)$ in terms of the Perey-Buck potential V(E) by means of the second order expansion

$$(3.28)\qquad \tilde{V}(E) = V(E) - \frac{\alpha}{4}(\underline{\nabla}^2V(E)) + \frac{\alpha^2}{32}(\underline{\nabla}^4V(E)) - \frac{\alpha^2}{2}\frac{(\underline{\nabla}V(E))^2}{1-\alpha V(E)} ,$$

which is deduced easily from Eqs. (3.3), (3.4) and (3.8). Neglecting some surface corrections of the third order, Eqs. (3.26) and (3.27) become

$$(3.29)\qquad \mathcal{V}_L(E) \simeq V(E) - \frac{\alpha}{4}(\underline{\nabla}^2V(E)) + \frac{\alpha^2}{32}(\underline{\nabla}^4V(E)) ,$$

$$(3.30)\qquad \phi \simeq \left(1-\alpha V(E) + \frac{\alpha^2}{4}(\underline{\nabla}^2V(E))\right)^{-1/2}\left(1+\frac{\alpha^2}{8}\frac{(\underline{\nabla}^2V(E))}{1-\alpha V(E)} + \frac{\alpha^2}{4}\frac{(\underline{\nabla}V(E))}{1-\alpha V(E)}\cdot\underline{\nabla}\right)\Psi_L ,$$

which must be compared with Eqs. (3.14) and (3.15).

Eq. (3.29) removes the last term of Eq. (3.14) and provides a second order surface correction which appears to be very slight.

A more interesting effect is observed in Eq. (3.30). In fact, the corrections to Eq. (3.15) regard also the nuclear interior. Here, we obtain

$$\phi \simeq \left(1-\alpha V(E)\right)^{-1/2}\Psi_L \quad , \tag{3.31}$$

which coincides exactly with the empirical formula found by Perey.[15)] Concerning the nuclear surface, we remark that the nonlocal wave function is affected also by an impulse-depending contribution. It should be interesting to evaluate numerically, by means of the wave functions obtained from local analyses, how much this effect contributes to the shape of ϕ when the scattering energy increases.

It remains to examine the solutions which may be lost when the Frahn-Lemmer potential is replaced by the local one.

About the scattering problem, no concrete complication arises. In fact, provided that it is everywhere $\tilde{V}(E) \neq \alpha^{-1}$ (as it is the case in practice), the equivalent local potential gives origin always to a scattering solution. Therefore, only the exceeding nonlocal solutions (i.e. those which are connected with a spurious state) can be lost. This was expected and does not compromise a complete description of the scattering.

On the contrary, more caution is due if one considers the bound state problem, which can be included in the above treatment without changing the formalism. In fact, there is no evidence at this stage for a one to one correspondence between local and nonlocal bound state solutions.

As regards the extension of the previous treatment to a higher order, in addition to the mathematical difficulties that are connected with the third order derivatives, some technical problems arise in order to remove the commutators. However, in the gaussian case an extension to the third order is possible if one uses a rather different method. This is based on a folded phase-conserving transformation as an intermediate step before expanding about Λ. In this way, a more accurate evaluation is obtained without increasing the order of the derivatives. Results for this case will be published likely in a near future.

4. A generalized Kisslinger potential

In the previous Section, the Kisslinger potential obtained from the Frahn-Lemmer model (see Eqs. (3.11) - (3.13))provides the first order nonlocal wave functions, whereas its local equivalent reproduces only their asymptotic behaviour. At this point an interesting question should be answered. Let us neglect the model and the related approximations; can an almost local potential yield exactly the wave functions of the theoretical optical potential (i.e. the physical ones)?

Indeed, in addition to a pure theoretical interest, the problem deserves also a practical one, since the numerical solution of an almost local Schrödinger equation is more simple and requires a less amount of computer time.

Really, the Kisslinger potential gives successful fits to experimental pion-nucleus data [9-13] even in an energy region where its theoretical derivation is not full justified. This fact too, although it does not support any conclusion about the physical wave functions, increases our interest in the almost local interaction and suggests an investigation in this direction.

For this purpose, we try to obtain some indications by starting from a more general approach to the Kisslinger potential that is founded on the precise requirement of a full equivalence to the nonlocal one. We emphasize that this approach is not suitable in order to calculate explicitely the Kisslinger potential. Nevertheless, this conceptual advantage is achieved: no convergence's problem and no derivative of order higher than the second are concerned.

Let us consider the Schrödinger equation for the theoretical optical potential:

(4.1) $$(\underline{\nabla}^2+E)\Psi(\underline{q},\underline{r}) = \int \mathcal{V}(E;\underline{r},\underline{r}')\Psi(\underline{q},\underline{r}')d\underline{r}' \quad , \; q^2 = E.$$

We require only the symmetry property, the rotational invariance and a short range of the interaction:

(4.2) $$\mathcal{V}(E;\underline{r},\underline{r}') = \mathcal{V}(E;\underline{r}',\underline{r}) = \mathcal{V}(E;r,r' \, , \frac{\underline{r}\cdot\underline{r}'}{rr'}) \quad ,$$

(4.3) $$|\,\mathcal{V}(E;\underline{r},\underline{r}')\,| \leqslant C(E)\exp\left(-\left(\frac{r+r'}{R}\right)\right) \quad .$$

Essentially, these conditions are also those assumed in Refs.17 to 19, where a rigorous scattering theory is developed. Indeed, the quoted papers concern only a real potential, but an extension to the complex case does not seem a difficult task.

For convenience, we denote by $\psi^+(\underline{q},\underline{r})$ the scattering solution. Moreover, by taking into account that the potential is parity-conserving, we consider also the solution

(4.4) $$\Psi^-(\underline{q},\underline{r}) \equiv \Psi^+(\underline{q},-\underline{r})$$

and then we define the following complex vectors:

(4.5) $$\underline{J}(\underline{q},\underline{r}) \equiv (2iq)^{-1}\left(\left(\Psi^-(\underline{q},\underline{r})\underline{\nabla}\Psi^+(\underline{q},\underline{r})\right)-\left(\Psi^+ \leftrightarrow \Psi^-\right)\right),$$

(4.6) $$\underline{\tilde{J}}(\underline{q},\underline{r}) \equiv E\underline{J}(\underline{q},\underline{r})+(2iq)^{-1}\left(\left((\underline{\nabla}^2\,\Psi^-(\underline{q},\underline{r}))\underline{\nabla}\Psi^+(\underline{q},\underline{r})\right)-\left(\Psi^+ \leftrightarrow \Psi^-\right)\right).$$

By means of the Schrödinger equation (4.1) one obtains two integral expressions for div $\underline{J}$ $(\underline{q},\underline{r})$ and $\underline{\tilde{J}}(\underline{q},\underline{r})$:

$$(4.7) \qquad \mathrm{div}\,\underline{J}(\underline{q},\underline{r})=(2iq)^{-1}\left(\left(\int \mathcal{V}(E;\underline{r},\underline{r}')\Psi^{-}(\underline{q},\underline{r})\Psi^{+}(\underline{q},\underline{r}')d\underline{r}'\right)-\left(\Psi^{+}_{\leftrightarrow}\Psi^{-}\right)\right),$$

$$(4.8) \qquad \underline{\tilde{J}}(\underline{q},\underline{r})=(2iq)^{-1}\left(\left(\int \mathcal{V}(E;\underline{r},\underline{r}')\Psi^{-}(\underline{q},\underline{r}')\underline{\nabla}\Psi^{+}(\underline{q},\underline{r})d\underline{r}'\right)-\left(\Psi^{+}_{\leftrightarrow}\Psi^{-}\right)\right) .$$

Starting from these last quantities, a full equivalent potential of the Kisslinger's type can be constructed easily, as it is precised by the following statement.

Provided that $\tilde{B}(\underline{q},\underline{r})$ is the solution of the equation

$$(4.9) \qquad \mathrm{div}\underline{J}(\underline{q},\underline{r}) = \underline{J}(\underline{q},\underline{r})\cdot\underline{\nabla}\,\tilde{B}(\underline{q},\underline{r}) \;,$$

which satisfies the boundary condition

$$(4.10) \qquad \lim_{r\to\infty} r\,\tilde{B}(\underline{q},\underline{r}) = 0 \;,$$

one can write the nonsymmetrical Kisslinger equation

$$(4.11) \qquad (\underline{\nabla}^2+E)\Psi^{\pm}(\underline{q},\underline{r}) = \left(\tilde{A}(\underline{q},\underline{r})+(\underline{\nabla}\tilde{B}(\underline{q},\underline{r}))\cdot\underline{\nabla}\right)\Psi^{\pm}(\underline{q},\underline{r})$$

or its symmetrical equivalent one

$$(4.12) \qquad (\underline{\nabla}^2+E)\Psi^{\pm}(\underline{q},\underline{r}) = \left(A(\underline{q},\underline{r})-\underline{\nabla}\cdot(B(\underline{q},\underline{r})\underline{\nabla})\right)\Psi^{\pm}(\underline{q},\underline{r}) \;,$$

where

$$(4.13) \qquad \tilde{A}(\underline{q},\underline{r}) = \frac{\underline{\tilde{J}}(\underline{q},\underline{r})\cdot\;\underline{\nabla}\tilde{B}(\underline{q},\underline{r})}{\mathrm{div}\;\underline{J}(\underline{q},\underline{r})} \;,$$

$$(4.14) \qquad A(\underline{q},\underline{r}) = \left(1-\exp\left(-\tilde{B}(\underline{q},\underline{r})\right)\right)E+\tilde{A}(\underline{q},\underline{r})\exp\left(-\tilde{B}(\underline{q},\underline{r})\right) \;,$$

$$(4.15) \qquad B(\underline{q},\underline{r}) = \exp\left(-\tilde{B}(\underline{q},\underline{r})\right)-1 \quad .$$

This statement follows directly from the identity

$$(4.16) \qquad \underline{J}(q,\underline{r})(\underline{\nabla}^2+E)\psi^{\pm}(q,\underline{r}) = \mathrm{div}\underline{J}(q,\underline{r})\underline{\nabla}\psi^{\pm}(q,\underline{r})+\underline{\tilde{J}}(q,\underline{r})\psi^{\pm}(q,\underline{r}) \quad ,$$

if one multiplies both sides by $\underline{\nabla}\tilde{B}(q,\underline{r})$.

As a consequence of the above definitions, the Kisslinger potential is parity-conserving. It is easy to recognize that the existence of the solution $\tilde{B}(q,\underline{r})$ of Eq. (4.9) is also a necessary condition in order that a full equivalent parity-conserving Kisslinger potential exists. Thus, no ambiguity in its deduction arises.

The boundary condition (4.10) is required in order that the potential decreases sufficiently fast at infinity. It is also a sufficient condition in order that a local phase-equivalent potential exists (see Sect. 1).

In some exceptional cases, a nonlocal potential may provide some solutions for which it is identically $\mathrm{div}\underline{J}(q,\underline{r})=0$. It follows that the equivalent potential is a local one, since from the identity (4.16) one obtains

$$(4.17) \qquad (\underline{\nabla}^2+E)\psi^{\pm}(q,\underline{r}) = \frac{\underline{J}(q,\underline{r})\cdot\underline{\tilde{J}}(q,\underline{r})}{(\underline{J}(q,\underline{r}))^2}\,\psi^{\pm}(q,\underline{r}) \; .$$

Now we examine briefly some general properties of the coefficients. Multiplying Eqs. (4.16) respectively by $(\underline{J}(q,\underline{r})\cdot\underline{\nabla}\psi^{\mp}(q,\underline{r}))\underline{\nabla}\tilde{B}(q,\underline{r})$ and subtracting the one from the other, we obtain

$$(4.18) \qquad \tilde{A}(q,\underline{r}) = \frac{\underline{J}(q,\underline{r})\cdot\underline{\tilde{J}}(q,\underline{r})}{(\underline{J}(q,\underline{r}))^2}$$

$$+ \frac{1}{2iq(\underline{J}(q,\underline{r}))^2}\Big(\big((\underline{J}(q,\underline{r})\cdot\underline{\nabla}\psi^{-}(q,\underline{r}))(\underline{\nabla}\tilde{B}(q,\underline{r})\cdot\underline{\nabla}\psi^{+}(q,\underline{r}))\big)-(\psi^{+}\leftrightarrow\psi^{-})\Big).$$

Eq. (4.18) shows that, with the exception at most of the zeros of

$(\underline{J}(\underline{q},\underline{r}))^2$, $\tilde{A}(\underline{q},\underline{r})$ is a regular function of $\underline{q}$ and $\underline{r}$ provided the same property is shared by the first $\underline{r}$-derivatives of $\tilde{B}(\underline{q},\underline{r})$.

About the coefficient $\tilde{B}(\underline{q},\underline{r})$, its properties of regularity can be drawn only from an investigation of Eq. (4.9). This in general is a difficult mathematical task since Eq. (4.9), owing to the complex character of its coefficients, is a system of two coupled real equations. The technical problem is that one deals with an elliptic system which degenerates in some particular regions, for instance on the boundary. At present, we have developed a complete treatment only for real coefficients. In this case, under general conditions, the existence and the uniqueness of the solution $\tilde{B}(\underline{q},\underline{r})$ have been proved. Furthermore, we have shown that $\tilde{B}(\underline{q},\underline{r})$ and its first derivatives are regular, provided it is everywhere $(\underline{J}(\underline{q},\underline{r}))^2 \neq 0$.

Apart from the incompleteness of the mathematical treatment of Eq. (4.9), it is evident that the properties of regularity of the Kisslinger potential depend upon the zeros of $(\underline{J}(\underline{q},\underline{r}))^2$. It is not clear yet when these occur in practice; at present the problem is investigated in a soluble model. We remark that the zeros of $(\underline{J}(\underline{q},\underline{r}))^2$ do not seem to be connected with the existence of spurious bound states. However, when these appear, another complication arises: the coefficients of the Kisslinger potential can be multivalued functions of $\underline{q}$ and $\underline{r}$.

When no spurious bound state is concerned, the scattering solution is unique and, hence, it shows a cylindrical symmetry around the scattering axis. Thus, the coefficients depend only upon q, r and the axial component z. The dependence upon q gives origin to a spurious energy dependence (as we see from Eqs. (4.7) and (4.8)) which reveals the non-causal character of the Kisslinger potential. A further and unexpected spurious effect is the dependence upon z, by which the potential shows only a cylindrical symmetry around the scattering axis.[x]

The breakdown of the causality property and of the spherical symmetry are two serious conceptual defects. They reveal that the Kisslinger potential, even if it provides the physical wave functions, is not the "true" potential.

Obviously, the deviation from the spherical symmetry is also a defect in view of the numerical complexity of a partial wave analysis, but this complication cannot be removed without changing the wave functions. Therefore, in principle, the following alternative occurs: either one uses a complicated potential which provides the physical wave functions or a spherical symmetric but only phase-equivalent potential must be employed.

From a more concrete point of view, the problem consists in singling out the cases in which the spherical symmetry is conserved approximately. Obviously, this occurs when the nonlocal potential has a sufficiently small range of nonlocality. Independently of this case, one can find an approximation which assures the spherical symmetry. For this purpose, we perform by means of the Legendre polynomials the following expansion of $\underline{J}(\underline{q},\underline{r})$ and $\underline{\tilde{J}}(\underline{q},\underline{r})$ over the variable $\cos\Theta \equiv \underline{q}.\underline{r}/qr$:

$$(4.19)\qquad \underline{J}(\underline{q},\underline{r})=\sum_{n=0}^{\infty}\left(f_{2n}(q,r)P_{2n}(\cos\Theta)\,\frac{\underline{q}}{q}+f_{2n+1}(q,r)P_{2n+1}(\cos\Theta)\,\frac{\underline{r}}{r}\right),$$

$$(4.20)\qquad \underline{\tilde{J}}(\underline{q},\underline{r})=\sum_{n=0}^{\infty}\left(\tilde{f}_{2n}(q,r)P_{2n}(\cos\Theta)\,\frac{\underline{q}}{q}+\tilde{f}_{2n+1}(q,r)P_{2n+1}(\cos\Theta)\,\frac{\underline{r}}{r}\right),$$

[x]The breakdown of the spherical symmetry affects also its phase-equivalent local potential (see Eq. (1.11)). We emphasize that this effect is different from the angular momentum dependence of the equivalent potentials obtained by other authors.[25-28] In fact these are effectively local in the radial coordinate but not in the angular ones.

where the symmetry of $\underline{J}(\underline{q},\underline{r})$ and $\tilde{\underline{J}}(\underline{q},\underline{r})$ for a space-inversion $\underline{r}\to-\underline{r}$ has been taken into account. If the above development can be truncated to the lowest Legendre polynomials by writing

(4.21) $$\underline{J}(\underline{q},\underline{r}) \simeq f_o(q,r)\,\frac{\underline{q}}{q} + f_1(q,r)\cos\theta\,\frac{\underline{r}}{r} \quad ,$$

(4.22) $$\tilde{\underline{J}}(\underline{q},\underline{r}) \simeq \tilde{f}_o(q,r)\,\frac{\underline{q}}{q} + \tilde{f}_1(q,r)\cos\theta\,\frac{\underline{r}}{r} \quad ,$$

one obtains the spherical symmetric expressions of the coefficients:

(4.23) $$\tilde{A}(q,r) \simeq \frac{\tilde{f}_o(q,r)+\tilde{f}_1(q,r)}{f_o(q,r)+f_1(q,r)} \quad ,$$

(4.24) $$\tilde{B}(q,r) \simeq \ln\left(r^2\left(f_o(q,r)+f_1(q,r)\right)\right)+2\int_r^\infty \frac{f_o(q,r')}{f_o(q,r')+f_1(q,r')}\,\frac{dr'}{r'} ,$$

and the related expressions for $A(q,r)$ and $B(q,r)$ (see Eqs. (4.14) and (4.15)).

From a qualitative point of view, this approximation can be justified in the limit cases of high and low energies. In fact in the high energy region, provided that the eikonal approximation is available, Eqs. (4.21) and (4.22) are obviously true. In this case the Kisslinger potential in also local (since it is div $\underline{J}(\underline{q},\underline{r})\simeq 0$) and coincides with the Perey-Buck potential. At low energy Eqs. (4.21) and (4.22) are justified if it is $qR\ll 1$ (where R is the range of the nonlocal potential, see Eq. (4.3)) and only s and p waves contribute to the scattering solution.

Thus, at low energy a spherical symmetric Kisslinger potential can be obtained even when the Fiedeldey approximation is not sufficiently accurate or when the nonlocal potential does not allow a Perey-Saxon expansion.[x] As regards the intermediate energy region, the deviation from the spherical symmetry is expected to be chiefly a surface effect, but only a numerical calculation of the nonlocal wave functions can give more details about this point.

---

[x]Work is now in progress in order to derive a Kisslinger potential from a finite rank nonlocal model.

ACKNOWLEDGMENTS

It is a pleasure to thank Profs. M.Bertero, G.Passatore and G.Talenti for helpful discussions.

REFERENCES

1) K.M.Watson, Phys.Rev. 89, 575 (1953); 105, 1388 (1957).
2) H.Feshbach, Ann.Rev. Nucl. Sci. 8, 49 (1958).
3) H.Feshbach, Ann. of Phys. 5, 357 (1958).
4) J.-P.Jeukenne, A.Lejeune and C.Mahaux, Physics Reports (1976).
5) G.Dillon and G.Passatore, Nucl.Phys. A114, 623 (1968).
6) G.Passatore, Nucl.Phys. A95, 694 (1967); A110, 91 (1968).
7) M.Bertero and G.Passatore, Z.Naturforsch. 28a, 519 (1973).
8) L.Kisslinger, Phys. Rev. 98, 761 (1955).
9) W.F.Baker, H.Byfield and J.Rainwater, Phys. Rev. 112,1773 (1958).
10) R.M.Edelstein, W.F.Baker and J.Rainwater, Phys. Rev. 122, 252 (1961).
11) E.H.Auerbach, D.M.Fleming and M.M.Sternheim, Phys. Rev. 162, 1683 (1967).
12) L.S.Kisslinger and F.Tabakin, Phys. Rev. C9, 188 (1974).
13) R.H.Landau, S.C.Phatak and F.Tabakin, Ann.of Phys.(N.Y.)78, 299

(1973).

14) H.Fiedeldey, Nucl.Phys. 77, 149 (1966).

15) F.G.Perey, "Direct interactions and nuclear reaction mechanisms" ed. by E.Clementel and C.Villi, Gordon and Breach, New York (1963).

16) R.G.Newton, Jour.Math. Phys. 1, 319 (1960).

17) M.Bertero, G.Talenti and G.A.Viano, Nuovo Cimento 46, 337 (1966).

18) M.Bertero, G.Talenti and G.A.Viano, Comm.Math. Phys. 6, 128 (1967).

19) M.Bertero, G.Talenti and G.A.Viano,Nucl.Phys. A113, 625 (1968); A115, 395 (1968).

20) M.Bertero and G.Dillon, Nuovo Cimento A2, 1024 (1971).

21) W.E.Frahn and R.H.Lemmer, Nuovo Cimento 5, 1564 (1957).

22) M.M.Giannini and G.Ricco, preprint Genova, November 1975.

23) F.G.Perey and B.Buck, Nucl. Phys. 32, 353 (1962).

24) F.G.Perey and D.S.Saxon, Phys. Lett. 10, 107 (1964).

25) H.Fiedeldey, Nucl. Phys. A96, 463 (1967).

26) M.Coz, A.D.Mackellar and L.G.Arnold, Ann. of Phys. 58, 504 (1970).

27) M.Coz, L.G.Arnold and A.D.Mackellar, Ann. of Phys. 59, 219 (1970).

28) F.Capuzzi, Nuovo Cimento A11, 801 (1972).

GREEN FUNCTION APPROACH TO SINGLE PARTICLE STATES IN NUCLEI

S. BOFFI

Istituto di Fisica Teorica, Università di Pavia, Pavia
Istituto Nazionale di Fisica Nucleare, Sezione di Pavia

Abstract. The Green function is introduced in the classical case of a damped-driven harmonic oscillator in order to understand its significance in a simple case. Transition to quantum mechanics allows to review the properties of the single particle Green function in many-body systems. Attention is then focussed on the spectral function and the definition of single particle energies. The connection between the energy sum rule for the ground state and information from quasi-free scattering is finally discussed. In particular, some uncertainties in the data analysis are pointed out, and the relationship between the validity of the energy sum rule and Koopmans theorem is presented.

1. Introduction

In the investigation of many-body systems it is convenient to assume that the degrees of freedom are labelled by quantum numbers characterizing a single particle (s.p.) state in some external (possibly self-consistent) average potential well. It is not necessary that this state corresponds exactly to a s.p. state: it may well describe a quasi-particle excitation in the sense of Landau theory of Fermi systems.

Here attention is devoted mainly to the properties of the s.p. Green function and its application to finite nuclei. Therefore, the reference many-body system is built with A nucleons. In the original shell model each nucleon is moving independently of the others in a common harmonic oscillator potential well with spring parameter $b = (\hbar/m\omega_o)^{1/2}$.

In the classical limit each nucleon can then be represented as a par ticle of mass m, oscillating with a frequency ω_o around the equilibrium position x = 0 of the centre of the well. Due to an external driving force, $F(t) = F_o \exp(i\omega t)$, the equation of motion of the classical oscillator,

$$\ddot{x} + \omega_o^2 x = \frac{F_o}{m} \exp(i\omega t) \quad , \tag{1}$$

has the following regime solution

$$x(t) = G^{(o)}(\omega)\, F_o \exp(i\omega t) \quad , \tag{2}$$

where the amplitude

$$G^{(o)}(\omega) = \frac{1}{m(\omega_o^2 - \omega^2)} \tag{3}$$

relates the eigenfrequency ω_o of the system to the character of the external perturbation (ω). Apart from unessential factors, $G^{(o)}$ is the inverse of (H-w), i.e. gives a resonance condition for the amplitude everywhere the applied energy w equals the eigenvalue of the system Hamiltonian H corresponding to the energy of an elementary excitation. $G^{(o)}(\omega)$ is called the Green function (or resolvent) of the system. Its Fourier transform $G^{(o)}(t)$ satisfies the following equation:

$$\left(\frac{d^2}{dt^2} + \omega_o^2\right) G^{(o)}(t) = \delta(t) \tag{4}$$

and represents the oscillation produced by a δ-type driving force: therefore $G^{(o)}$ is also called the response function.

In the presence of an interaction between elementary excitations, energy transfer is allowed to other degrees of freedom: this introduces a finite life-time of the individual elementary excitation, i.e. a damping term in the equation of motion of a single oscillator:

$$\ddot{x} + \gamma x + \omega_o^2 x = \frac{F_o}{m} \exp(i\omega t) \quad , \tag{5}$$

whose solution is

$$x(t) = G(\omega)\, F_o \exp(i\omega t) \quad , \tag{6}$$

where

$$G(\omega) = \frac{1}{m(\omega_o^2 - \omega^2) - M(\omega)} \quad . \tag{7}$$

The poles of the Green function for the interacting system $G(\omega)$ are complex and are obtained in terms of the self-energy

$$M(\omega) = -i\gamma m\omega \tag{8}$$

as solutions of the following equation

$$m(\omega_o^2 - \omega^2) - M(\omega) = 0. \tag{9}$$

To characterize the elementary excitations it is useful to consider the spectral function $S(\omega)$ defined as

$$\begin{aligned} S(\omega) &= -2 \operatorname{Im} G(\omega) \\ &= - \frac{2 \operatorname{Im} M(\omega)}{m^2(\omega_o^2 - \omega^2)^2 - \left[\operatorname{Im} M(\omega)\right]^2} . \end{aligned} \tag{10}$$

It reduces to a delta-function peaked at the eigenfrequency ω_o, if $\gamma \to 0$. For $\gamma \neq 0$ the frequency dependent width of the peak in the spectral function is related to the life-time of the elementary excitation and contains, through γ, all information concerning the interaction between different excitation modes. As long as $\operatorname{Im} M(\omega)$ is small, i.e. the life-time is large, the s.p. description is useful. Otherwise collective effects are important, and the analysis of the system involves other degrees of freedom.

It is also interesting to notice an algebraic relation between the Green function $G^{(o)}$ for the interaction-free case and the Green function G for the interacting system, which is known as the Dyson equation:

(11) $$G(\omega) = G^{(o)}(\omega) + G^{(o)}(\omega)\, M(\omega) G(\omega) \quad .$$

If γ is small, a perturbative solution of Eq. (11) is possible:

(12) $$G(\omega) = G^{(o)}(\omega) + G^{(o)}(\omega)\, M(\omega) G^{(o)}(\omega) + + G^{(o)}(\omega) M(\omega) G^{(o)}(\omega) M(\omega) G^{(o)}(\omega) + \ldots \quad ,$$

which can be represented in terms of graphs as

(13) $G(\omega) =$ + + + ...

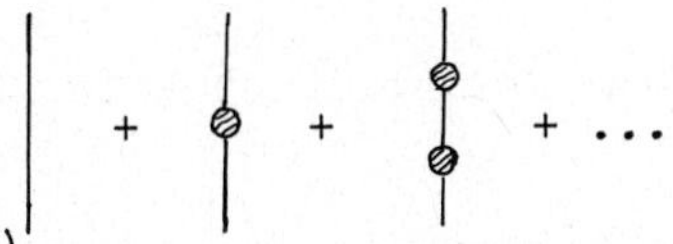

Therefore, the knowledge of $G^{(o)}$ and of the self-energy $M(\omega)$ of individual excitations determines $G(\omega)$. The ω-dependence of M may also be much more complicated than in Eq. (8). This would result, however, only in a more complicated structure of the singularities of G determined by Eq. (9) (see Sect. 2).

In the next Section this classical picture is substituted by the necessary quantum mechanical generalization, and the s.p. Green function is discussed; its relation with the energy sum rule for the ground state is presented in Sect.3, where also a review is presented of the theoretical and experimental definitions of s.p. energies. Information from knock-out reactions about the hole part of the s.p. Green function is analyzed in Sect. 4.

2. The single particle Green function

The time dependent s.p. Green function for a system, whose true ground state is $|\psi\rangle$, is defined as [1,2]

(14) $$G_{\alpha\beta}(t) = -i\langle\Psi| T \left\{ a_\alpha(t) a_\beta^+(o) \right\} |\Psi\rangle$$

where $a_\alpha^+(t)$ is an operator in the Heisenberg representation which creates a particle at time t with quantum numbers labelled by α:

(15) $$a_\alpha^+(t) = e^{iHt} a_\alpha^+ e^{-iHt} \quad .$$

Due to the Wick's time ordering operator T, Eq. (14) reads also as

(16a), (16b) $$G_{\alpha\beta}(t) = \begin{cases} -i\langle\Psi|a_\alpha e^{-i(H-E_o)t}\, a_\beta^+|\Psi\rangle, & t>o \quad , \\ i\langle\Psi|a_\beta^+\, e^{i(H-E_o)t}\, a_\alpha|\Psi\rangle, & t<o \quad . \end{cases}$$

In Eq. (16) use has been made of the fact that E_o is the ground state energy corresponding to $|\Psi\rangle$ and, as in Eq. (14), the Hamiltonian H is time independent.

The process involved in Eq. (16a), e.g. for $\alpha=\beta$, is as follows:

i) at time t=0, a particle is created, and $a_\alpha^+|\Psi\rangle$ represents the new state of the system with one added particle;

ii) $\exp(-iHt)a_\alpha^+|\Psi\rangle$ is the time displaced state of the system at time t, when

iii) the same particle is removed;

iv) the new state reached by the system at time t, $a_\alpha \exp(-iHt)a_\alpha^+|\Psi\rangle$, is overlapped with the time displaced state of the unperturbed system, $\exp(-iE_o t)|\Psi\rangle$.

Therefore, the diagonal part of Eq. (16a) is proportional to the probability amplitude of the propagation of a particle for t>0. Similarly, Eq. (16b) is related to the probability for first removing and then adding a particle (for t<0), i.e. for the propagation of a hole for t< 0.

In a non-interacting system

(17) $$G_{\alpha\beta}^{(o)}(t) = \pm\, i\delta_{\alpha\beta}\, \exp(-i\varepsilon_\alpha t) \quad , \quad t \gtrless 0 \quad ,$$

where ε_α is the single particle (hole) energy, i.e.

$$(18)\qquad \begin{array}{ll} \varepsilon_\alpha \lesssim \varepsilon_F & \text{for holes,} \\ \varepsilon_\alpha > \varepsilon_F & \text{for particles,} \end{array}$$

and ε_F is the Fermi level energy.

In Fourier space one has

$$(19)\qquad G_{\alpha\beta}(w) = \int G_{\alpha\beta}(t)\, e^{iwt}\, dt = G^p_{\alpha\beta}(w) + G^h_{\alpha\beta}(w) \quad ,$$

where the two contributions in Eq. (19) come from the two possibilities ($t \gtrless 0$) in Eq. (16):

$$(20)\qquad G^p_{\alpha\beta}(w) = \langle\Psi| a_\alpha \frac{1}{w+E_o-H+i\delta} a^+_\beta |\Psi\rangle \quad ,$$

$$(21)\qquad G^h_{\alpha\beta}(w) = \langle\Psi| a^+_\beta \frac{1}{w-E_o+H-i\delta} a_\alpha |\Psi\rangle \quad .$$

Eqs. (20) and (21) indicate that these amplitudes are governed by the energies of the single particle excitations as in the case of the classical harmonic oscillator of Sect. 1. In fact, for the interaction-free case the Fourier transform of Eq. (17) is

$$(22)\qquad G^{(o)}_{\alpha\beta}(w) = \frac{\delta_{\alpha\beta}}{w-\varepsilon_\alpha+i\delta} \quad , \quad \varepsilon_\alpha \gtrless \varepsilon_F \quad ,$$

an expression which is similar to Eq. (3).

The Green function for the interacting system can be obtained from $G^{(o)}$ via the Dyson equation (11). The Green function $G(w)$ is analytic in the complex w-plane except along the real axis where poles and branch cuts appear. The singularities of $G(w)$ can be calculated in terms of the self-energy $M(w)$. The poles are determined by the equation

$$(23)\qquad (w-\varepsilon_\alpha)\delta_{\alpha\beta} - M_{\alpha\beta}(w) = 0 \quad ,$$

and the continuum part consists of peaks located at energies which are solutions of

(24) $$(w - \varepsilon_\alpha)\delta_{\alpha\beta} - \mathrm{Re}\, M_{\alpha\beta}(w) = 0 \quad ,$$

with a width determined by $\mathrm{Im}\, M_{\alpha\beta}(w)$.

The analytic properties of the Green function are better investigated if one defines the spectral function for particles and holes as follows:

(25) $$S^p_{\alpha\beta}(w) = \langle\Psi| a_\alpha \; \delta(w+E_o-H) a^+_\beta \; |\Psi\rangle$$

$$= \sum_m \langle\Psi| a_\alpha |m\rangle \; \langle m| a^+_\beta |\Psi\rangle \; \delta(w-w_m) \quad ,$$

where $|m\rangle$ is an A+1 particle state,

(26) $$H|m\rangle = E'_m|m\rangle$$

(27) $$w_m = E'_m - E_o \quad ,$$

and

(28) $$S^h_{\alpha\beta}(w) = \langle\Psi| a^+_\beta \delta(w-E_o+H) a_\beta |\Psi\rangle$$

$$= \sum_j \langle\Psi| a^+_\beta |j\rangle\langle j| a_\alpha |\Psi\rangle \; \delta(w-w_j) \quad ,$$

where $|j\rangle$ is an A-1 particle state,

(29) $$H|j\rangle = E''_j \; | j\rangle \quad ,$$

(30) $$w_j = E_o - E''_j \quad .$$

Therefore, $S^p_{\alpha\alpha}(\omega)$ $(S^h_{\alpha\alpha}(\omega))$ measures the joint probability of creating (removing) a particle with quantum numbers α and of finding the resulting system with A+1 (A-1) particles in a state with energy $E'_m(E''_j)$.

The spectral function is defined only along the real axis of the complex w-plane, everywhere the energy delta function is satisfied both for bound (Eq. 23) and continuum (Eq. (24) states (Fig. 1).

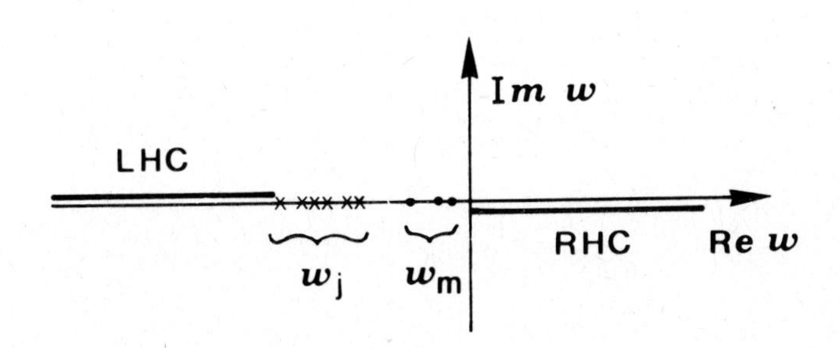

Fig. 1. The analytic structure of the single particle Green function in the w-plane (adapted from Ref. 3).

Alternatively, the spectral function can be conceived as the discontinuity of the s.p. Green function across the real w-axis (as in Eq.(10))

$$S_{\alpha\beta}(w) = S^{p}_{\alpha\beta}(w) + S^{h}_{\alpha\beta}(w) = \frac{1}{2\pi i}\{G_{\alpha\beta}(w-i\delta) - G_{\alpha\beta}(w+i\delta)\} . \tag{31}$$

Hence, a dispersion relation connects G and S:

$$G^{p}_{\alpha\beta}(w) = \sum_{m} \frac{S^{p}_{\alpha\beta}(w_m)}{w-w_m+i\delta} + \int_{RHC} dz \frac{S^{p}_{\alpha\beta}(z)}{w-z+i\delta} , \tag{32}$$

$$G^{h}_{\alpha\beta}(w) = \sum_{j} \frac{S^{h}_{\alpha\beta}(w)}{w-w_j-i\delta} + \int_{LHC} dz \frac{S^{h}_{\alpha\beta}(z)}{w-z-i\delta} \quad . \tag{33}$$

This representation for the Green function (where the integrals are performed along the right (R) or the left (L) hand cut of Fig. 1) clearly indicates that G^p and G^h are different parts of the total Green function G, which are not connected by analytic continuation. Therefore, the self-energy M for hole states cannot be regarded as the analytic continuation of the self-energy for particle scattering. Only the energy independent (Hartree-Fock) part of the hole and particle potentials can be identified within 1/A effects. [3)]

For scattering problems, $M_{\alpha\beta}(w)$ can be identified with the generalized optical model potential. [4)] Its calculation and discussion are presented in other talks during this meeting [5)] and will not be dealt with here. This kind of problem is concerned with the particle part of the Green function, whereas here the interest is mainly in the hole part.

In Fig. 2 the complicated structure is shown of the hole part of the diagonal spectral function in a typical case: [6)]

$$S^{h}_{\alpha\alpha}(w) = \sum_{i} z_{\alpha_i}\,\delta(w-w_{\alpha_i}) + \frac{1}{\pi}\,\frac{\mathrm{Im}M_{\alpha\alpha}(w)\;\Theta(w_o-w)}{\left(w-\varepsilon_\alpha-\mathrm{Re}M_{\alpha\alpha}(w)\right)^2+\left(\mathrm{Im}M_{\alpha\alpha}(w)\right)^2} \quad . \tag{34}$$

Here, w_o is the threshold of the continuum part of the spectrum for particle emission, and z_{α_i} is the residue of the Green function for the pole w_{α_i}:

$$z_{\alpha_i} = \left(1 - \frac{d}{dw} M_{\alpha\alpha}(w)\right)^{-1}_{w=w_{\alpha_i}} \quad . \tag{35}$$

It appears that the hole part of the spectral function has many interconnected peakes and resonances due to the w-dependence of M(w) even if the hole structure of the considered nucleus is assumed simple.

This fact shows the intimate relationship between the s.p. description of the (A-1) -particle system and the A-particle system.

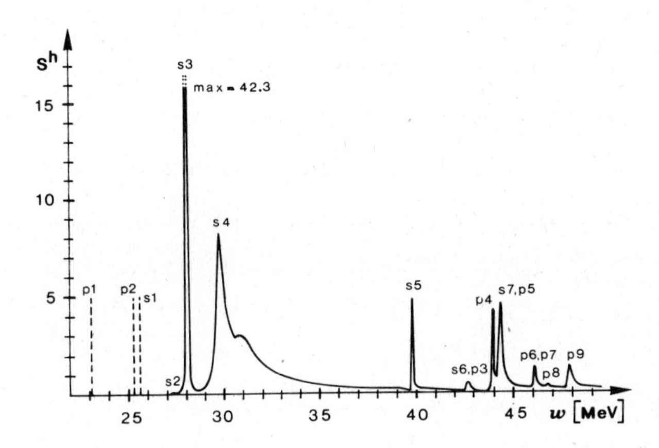

Fig. 2. The hole part of the spectral function of ^{12}C. The bound state peaks and resonances are labelled according to their spin and parities by $s(J^{\pi} = 1/2^{+})$ and $p(J^{\pi}=3/2^{-})$. The numbering serves to identify the solutions of Eq. (24) (adapted from Ref. 6).

When dealing with an extended system, where translational invariance holds, the convenient set of quantum numbers is represented by the momentum $\vec{p}$ of the particles, besides their spin. Then all the quantities defined in this section are diagonal in $\vec{p}$, and the formalism simplifies. However, in a finite system such as the nucleus, s.p. states are not eigenstates of the momentum. If the $\vec{p}$-representation is used, all quantities are not diagonal, just as in general for the α-represen tation. This fact has a great importance and its consequences are discussed in Sect. 4.

3. Single particle energies and the energy sum rule

In this section the discussion will be confined to hole states in nuclei. The pick-up process, as e.g. the (p,d) reaction, and the quasi-free (e,e'p) and (p,2p) reactions are a very useful tool for studying s.p. state of a nucleus. The angular distributions of products corresponding to different states of the final residual nucleus determine the appropriate quantum numbers α of the removed particle. The cross section for excitations of these final states gives a measure of how much the state can be described in terms of s.p. excitations, i.e. in terms of a level in an independent particle model.

However, one must distinguish between fast and slow reactions. In a fast reaction there are often several levels of the residual nucleus with the same spin and parity in the same energy region, which are excited to their maximal strength. Then the independent particle model level is distributed among several nuclear levels. In a slow reaction, time is given to the system to possibly relax to a final state involving only the configuration with the lowest missing energy.
Therefore, from an experimental point of view it may be convenient to define [7,8] an observable separation energy $\varepsilon_\alpha(\tau)$ which depends on the removal time τ:

(36) $$\varepsilon_\alpha(\tau) = \sum_j \sigma_{\alpha_j}(\tau)\,(E_o - E''_j) \quad .$$

In a sudden removal, $\tau\to o$

(37) $$\sigma_{\alpha_j}(0) = |\langle j|a_\alpha|\Psi\rangle|^2 / \sum_j |\langle j|a_\alpha|\Psi\rangle|^2 \quad ,$$

and $\varepsilon_\alpha(0)$ appears as the center fo gravity (centroid) of the group of nuclear levels involved, weighting each level in proportion to how strongly it is excited in the reaction. The weights, i.e. the spectroscopic factors $\sigma_{\alpha_j}(0)$, are interpreted as the probability that the level j of the residual nucleus looks like the target nucleus minus a single nucleon in the orbital α. Therefore, the quantity

$$\varepsilon_\alpha^M \equiv \varepsilon_\alpha(0) \tag{38}$$

is called the mean removal energy. It can be obtained from quasi-free knock-out reactions, as (e, e'p) and (p,2p).

Conversely, in the adiabatic limit, $\tau\to\infty$,

$$\sigma_{\alpha_j}(\infty) = \delta_{jj_o} \quad , \tag{39}$$

where j_o labels the state of the residual nucleus which has no extra excitation besides the one hole in the orbital α. Accordingly,

$$\varepsilon_\alpha^S \equiv \varepsilon_\alpha(\infty) = E_o - E''_{j_o} \tag{40}$$

is defined as the separation energy.

Large differences are observed between ε_α^S and ε_α^M (up to some MeV). The positive quantity

$$\Delta_\alpha = \varepsilon_\alpha^S - \varepsilon_\alpha^M \tag{41}$$

is called rearrangement energy, as it arises from an orbital rearrangement inside the nucleus after the sudden removal of one particle.

All the independent particle models aim at the description of s.p. states possibly in a self-consistent way. According to Brandow,[9,10] in nuclear matter as well as in finite nuclei, a convenient way of treating the self-energy M is to use the linked cluster theory and to consider all the insertions into particle or hole lines in the diagrams.

An insertion is part of a diagram, connected to the rest by two line segments, and has the dimension of energy. Insertions may be summed to produce a contribution $M'_{\beta\gamma}(w)$, where α and β label the external line segments and w is an energy parameter. The idea in self-consistent theories is to sum large classes of these insertions in such a way that the effect of $M'_{\beta\gamma}(w)$ is reproduced by a s.p. potential. One can divide the insertions into two classes, denoted by M^{on} and M^{off}, referring to on- and off-energy shell insertions. Brandow has shown that the $M^{on}(w)$ insertions are on-energy shell in the sense that the value of w is determined only by the energies of the external β,γ line segments and not by the rest of the larger diagram.

The subset of so-called irreducible on-energy shell insertions defines the s.p. potential $U_{\beta\gamma}(w)$. The self-consistent s.p. potential can be chosen in such a way that

$$U_{\beta\gamma}(\varepsilon_\alpha) = 0 \quad , \tag{42}$$

which in lowest order reduces to Hartree-Fock equations.

According to Koopmans theorem,[11] if alterations in the core orbitals can be neglected when a nucleon is added or removed, i.e. if orbital rearrangement is expected to be small, then the self-consistent s.p. energies are identical with the separation energies:

$$\varepsilon_\alpha = \varepsilon_\alpha^S \quad . \tag{43}$$

This identification, well known in the case of the Hartree-Fock approximation, is possible for any self-consistent variant of the independent particle model, i.e., in particular for Brueckner-Hartree-Fock, Renormalized BHF,[12] and Density Dependent HF.[13]

Furthermore, it has been shown by Koltun [14] that for normally occupied self-consistent orbits, the choice (42) produces self-consistent orbital energies identical with the mean removal energies:

(44) $$\varepsilon_\alpha = \varepsilon_\alpha^M \quad , \quad \alpha < F \quad .$$

Therefore, for any self-consistent independent particle model

(45) $$\Delta_\alpha = 0 \quad .$$

An indication of how large is the contribution of the rearrangement energy can be obtained through the use of the energy sum rule for the ground state:[1,2]

(46) $$E_0 = \frac{1}{2} \; Tr(T\rho) + \frac{1}{2} \lim_{t \to 0-} \sum_\alpha \frac{d}{dt} G_{\alpha\alpha}(t) \quad ,$$

where T is the kinetic energy, and ρ is the one-body density matrix:

(47) $$\rho_{\alpha\beta} = -i \lim_{t \to 0-} G_{\alpha\beta}(t) \quad .$$

Eq. (46) is exact if at most two-body forces are present.

By selecting, e.g. the natural orbital basis for s.p. states,

(48) $$\rho_{\alpha\beta} = n_\alpha \delta_{\alpha\beta} \quad , \quad 0 \leqslant n_\alpha \leqslant 1 \quad ,$$

(49) $$\begin{aligned} n_\alpha &= \langle\Psi| a_\alpha^+ a_\alpha |\Psi\rangle \\ &= \int S_{\alpha\alpha}^h(w)\,dw \quad . \end{aligned}$$

Moreover, from Eqs. (16b), (28), (37), (38)

(50) $$\begin{aligned} \lim_{t \to 0-} \frac{d}{dt} G_{\alpha\alpha}(t) &= \sum_j |\langle j| a_\alpha |\Psi\rangle|^2 w_j \\ &= \int w \; S_{\alpha\alpha}^h(w)\,dw \\ &= n_\alpha \varepsilon_\alpha^M \end{aligned}$$

Therefore

(51) $$E_0 = \frac{1}{2} \; Tr(T\rho) + \frac{1}{2} \sum_\alpha n_\alpha \varepsilon_\alpha^M \quad .$$

Eq. (51) allows to correlate the ground state energy to the experimental mean removal energies, if knowledge only of the true one-body density matrix is achieved. This method has been applied [15,17] with a realistic ρ fitting the elastic electron scattering data of closed shell spherical nuclei. At that time poor knowledge of ε^M was possible; therefore, ε^S was used so that the difference between the experimental and the calculated binding energy gives then an estimate of the rearrangement energy. In the case of ^{4}He (Ref.15), no negative values for E_o are possible for any realistic ρ. This fact was interpreted as an indication of large differences between mean removal energies and separation energies, but also as due to a possible overestimation of degrees of freedom, because the four particle ^{4}He system has in fact three degrees of freedom after removal of centre of mass motion.

In Table I results are shown for the average kinetic energy and rearrangement energy per particle with different one-body density matrices. It is remarkable the fact that realistic densities lower the kinetic energy with respect to pure HF calculations, thus supporting the idea [17] that the best independent particle model should be chosen according to the best-density criterion instead of the best-energy (HF) or the best overlap criterion (BHF, DDHF).

TABLE I

Average kinetic energy $\overline{T}$ and rearrangement energy $\overline{\Delta}$ per nucleon (in MeV) for ^{16}O and ^{40}Ca with several models. ρ_{SD} is the best idem-potent density matrix approximating the realistic ρ; ρ_{HF} is obtained in the Hartree-Fock approximation with the effective force of Ref.18; ρ_{DDHF} corresponds to the density-dependent Hartree-Fock of Ref. 19.

	^{16}O		^{40}Ca	
	$\overline{T}$	$\overline{\Delta}$	$\overline{T}$	$\overline{\Delta}$
ρ	15.07	2.72	18.81	5.45
ρ_{SD}	15.79	3.03	17.68	4.74
ρ_{HF}	18.54	4.39	26.17	8.94
ρ_{DDHF}	14.41	6.20	16.25	8.12

4. Knock-out reactions and spectral density function

In this section the relationship between knock-out reactions and the energy sum rule is considered by means of a discussion of the spectral density function used in the data analysis. Quasi-free scattering (Fig. 3) can be described in distorted wave impulse approximation (DWIA) with the following formula for the differential cross section:[3)]

$$\frac{d^6\sigma}{dE_{k'}\, d\Omega_{k'}\, dE_{p'}\, d\Omega_{p'}} = \tag{52}$$

$$= g\int d\vec{q}\; d\vec{q}\,'\; \hat{M}(\vec{k}'\vec{p}\,';\; \vec{k}\vec{q})\, S^{h}_{\vec{q}\vec{q}\,'}(w)\, \hat{M}^{+}(\vec{k}'\vec{p}\,';\vec{k}\vec{q}\,') \quad ,$$

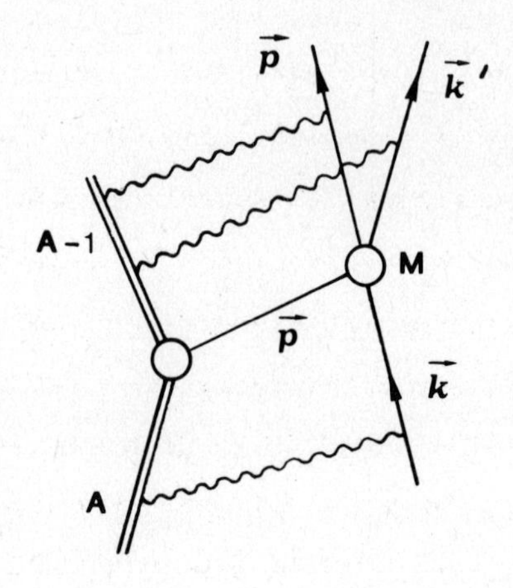

Fig. 3. The quasi-free knock-out process. The wavy lines indicate optical model interactions.

where g is a kinematical factor, and, e.g. for (e,e'p) reactions neglecting Coulomb distortion of electrons,

$$\hat{M}(\vec{k}'\vec{p}';\vec{k}\vec{q}) = \phi_{\vec{p}'}^{(-)}(\vec{k}+\vec{q}-\vec{k}')M(\vec{k}',\vec{k}+\vec{q}-\vec{k}';\vec{k}\vec{q}) \quad . \tag{53}$$

The quantity M is a half-off-energy shell two-body scattering matrix element, and $\phi^{(-)}$ is the proton outgoing distorted wave function in the momentum representation.

Eq. (52) can be simplified if the plane wave impulse approximation

(PWIA) holds:

$$\frac{d^6\sigma}{dE_{k'}d\Omega_{k'}dE_{p'}d\Omega_{p'}} = g|M(\vec{k}'\vec{p}';\vec{k}\vec{p})|^2 \; S^h_{\vec{p}\vec{p}}(w) \quad . \tag{54}$$

In this case only the diagonal part (in the momentum representation) of the spectral function is involved. Also, Eq. (52) essentially reduces to Eq. (54) if one is allowed to substitute the distorted wave function of the outgoing particle in Eq. (53) by a delta function times a reduction factor. This is often the case when the final state interaction is not so much to alter the momentum of the particle but to reduce its corrent.[20,21] However, this is clearly a rough simplification in the general case, and Eq. (52) should be preferred even for a small distortion, because it involves also the nondiagonal part of the spectral density whose contribution may be important.

Recent investigations, among others, in Tokyo [22,23] and particularly in Saclay [24,25] have shown the possibility of accurate experiments whose appropriate interpretation may give an experimental spectral function to be compared with the theoretical one.

Also, a phenomenological analysis proposed to satisfy the energy sum rule,[26] when applied to the recent data, seems to contradict it and to underestimate the total number of protons in the target.[25] More specifically, if one defines by E_z/Z the observed binding energy per proton, the energy sum rule (46) (corrected for centre of mass motion) reads as

$$\Delta = \frac{E_z}{Z} - \frac{1}{2}\left(\frac{A-2}{A-1}\,T_m + E_m\right) = 0 \quad , \tag{55}$$

where

$$T_m = N^{-1}\int dw \int d\vec{p}\,\frac{p^2}{2m}\,S^h_{\vec{p}\vec{p}}(w) \quad , \tag{56}$$

(57) $$E_m = N^{-1}\int dw \int d\vec{p}\; w\; S^h_{\vec{p}\vec{p}}(w) \quad ,$$

(58) $$N = \int dw \int d\vec{p}\; S^h_{\vec{p}\vec{p}}(w) \quad .$$

The experimental spectral density inserted in Eqs. (56) - (58) gives a non vanishing Δ and a value for N smaller than Z (Ref. 25). Only for ^{40}Ca, even with $N<Z$, Δ is nearly zero (Table II). This has been widely interpreted as a break-down of the independent particle model and as an indication of possible three-body forces inside the nucleus.[27) However, a careful consideration of all the uncertainties of data analysis along the lines below will show that these arguments are without justification at present.[28)

The uncertainty of data analysis can be summarized as follows.

1. Radiative corrections were calculated [24,25) and shown not to alter the shape but to reduce the size of the cross section. Therefore they were neglected in the analysis. This introduces an error of about 10% in the results.[25)

2. Coulomb distortion of electrons was neglected. This has been estimated to another 5% error. [25)

3. Statistical errors are at best between 3 and 5% [24,25) but may also be larger. [22,23).

TABLE II

Experimental binding energy E_z/Z, calculated average kinetic energy T_m and removal energy E_m per nucleon and failure Δ of the energy sum rule (in MeV). N is the total number of protons calculated from the experimental spectral function.[25)]

	E_z/Z	T_m	E_m	Δ	N
^{12}C	-6.93	16.9	23.4	-2.9 ± 0.5	3.5
^{28}Si	-6.84	17.0	24.5	-2.8 ± 0.6	9.7
^{40}Ca	-6.51	16.6	27.8	-0.7 ± 0.5	16.2
^{58}Ni	-6.95	18.7	24.8	-3.8 ± 0.7	26.2

Therefore, in the most favorable situation, combining 1. to 3. a possible error of 20% is already present in the data used to extract the experimental spectral function.[24)]

4. The cross section was never considered in the correct DWIA, Eq. (52), but a combination of PWIA plus a distorted momentum distribution has been always used.[20,23,25)]

This means that in Eq. (54) the spectral function was parametrized as follows

$$S^h_{\vec{p}\vec{p}}(w) = \sum_{\alpha} S^h_{\alpha\alpha}(w)\,|\phi_\alpha^{(-)}(\vec{p})|^2 \quad , \tag{59}$$

where $|\phi_\alpha^{(-)}(\vec{p})|^2$ is the distorted momentum distribution of the ejected proton inside the target nucleus.

In the correct DWIA formula, the distortion is included in the two-body scattering matrix $\tilde{M}$,not in S^h (see Eq. (53). Using a Lorentzian form for $\phi^{(-)}$ in Eq. (53) one can estimate to a few percent the error due to substituting M and Eq. (59) to $\tilde{M}$ and Eq. (52); but the correction is $\vec{p}$-dependent, thus involving both the diagonal and the nondiagonal parts of S^h with $\vec{p}$-dependent weights.

5. The spectral function, as shown e.g. in Fig. 2, is a quite complicated object which cannot be approximated even in PWIA by the simple formula (59), where each α is confined in practice to a particular energy range: peaks corresponding to different α appear in the same energy region thus making the deconvolution of the sum in Eq. (59) much more difficult.

6. The uncertainties in the imaginary part of the optical model potential used to calculate the distorted momentum distribution may have a rather larger effect, as already said in Ref. 25. This is a crucial point, because the corresponding error is unpredictable.

Therefore, combining 4. to 6. it is clear that the reliability of the experimentally obtained spectral function is already rather poor for a rigorous check of the energy sum rule.

7. A complete DWIA calculation, as in Eq. (52) involves the full spectral function and not only its diagonal part. Therefore, even in the simple scheme of Eq. (59), nondiagonal contributions $S^h_{\alpha\beta}(w)$ appear in the orbital analysis. These result in an additive term in the kinetic energy even in the case of zero spin target nuclei.

This term, of the form

$$\frac{1}{2N} \frac{A-2}{A-1} \sum_{\alpha \neq \beta} \rho_{\alpha\beta} T_{\alpha\beta} \qquad (60)$$

is by no means negligible.

8. Let us assume that from experiment one can determine $S^h_{\alpha\alpha}(w)$ to

obtain from Eq. (49) the occupation probability n_α of the orbital α within an error δn_α of the first order. This will in general produce a first order error in N and Δ, too, as shown in Table II for all nuclei but ^{40}Ca. In a forthcoming paper, [28] where all these points are discussed with greater details, it is shown that if one assumes also a diagonal ρ confined to occupied states, i.e.

$$\rho_{\alpha\beta} = n_\alpha \delta_{\alpha\beta} \quad , \qquad \alpha \lessgtr F \quad ,$$

then to a first order error in N corresponds a second order error in Δ. This situation arises everywhere the residual nucleus can be well described in terms of the target nucleus with one hole, as for the doubly magic ^{40}Ca nucleus. On the contrary, nonclosed shell nuclei, well known as deformed nuclei, are poorly reproduced by a diagonal spectral function, or in other words hardly satisfy Koopmans theorem. Indeed, because the definition of the spectral function involves both the target and the residual nucleus, it is much more probable that the same orbital basis is suited for both nuclei if the target one is magic (and spherical) than if it is deformed. In this frame the ^{40}Ca case is no more puzzling, as in Refs. 25 and 27.

The present discussion also emphasizes that some caution must be exercised, before talking of failure of the independent particle model or of contribution of three-body forces. The situation is rather encouraging a better analysis of the recent very accurate experiments in the frame of the Green function approach and the DWIA.

References

1) V.M. Galitski and A.B.Migdal, Sov. Phys. JETP 7, 96 (1958).

2) D.J. Thouless, The Quantum Mechanics of Many-body Systems, Academic Press, New York and London, 1961.

3) D.H.E. Gross and R.Lipperheide, Nuclear Physics A150, 449 (1970).

4) J.S.Bell and E.J.Squires, Phys. Rev. Lett. 3, 96 (1959).

5) J.-P.Jeukenne,A.Lejeune, C.Mahaux, Many-Body theory of the optical model potential, this Conference.

6) U.Wille and R.Lipperheide, Nuclear Physics A189, 113, (1972).

7) H.W.Meldner and J.D.Perez, Phys. Rev. A4, 1388 (1971).

8) K.A.Brueckner, A.W.Meldner and J.D.Perez, Phys. Rev. C6, 773 (1972).

9) B.H.Brandow, Phys. Rev. 152, 863 (1966).

10) B.H.Brandow, Ann. of Physics (N.Y.) 57, 214 (1970).

11) T.Koompans, Physica 1, 104 (1934).

12) R.L.Becker and M.R.Patterson, Nuclear Physics A178, 88 (1971).

13) G.Ripka, Saclay Rep. DPh-T/69-54 (1969).

14) D.S.Koltun, Phys. Rev. C9, 484 (1974).

15) S.Boffi, Nuovo Cimento Lettere 1, 931 (1971).

16) S.Boffi and F.D.Pacati, Nuclear Physics A204, 485 (1973).

17) S.Boffi and F.D.Pacati, Istituto Lombardo (Rend. Sci., Milano) A107, 321 (1973) .

18) C.M.Shakin, Y.R. Waghmare, M.Tomaselli and M.H.Hull, Phys. Rev. 161, 1015 (1967).

19) J.W.Negele, Phys. Rev. C1, 1260 (1970).

20) G.Jacob and Th. A.Maris, Rev.Mod.Phys. 38, 121 (1966); 45, 6 (1973).

21) S.Boffi, M.Bouten, C.Ciofi degli Atti and J.Sawicki, Nuclear Physics A120, 135 (1968).

22) H.Hiramatsu, T.Kamae, H.Muramatsu, K.Nakamura, N.Izutsu and Y.Watase, Physics Lett. 44B, 50 (1973).

23) K.Nakamura, S.Hiramatsu, T.Kamae, H.Muramatsu, N.Izutsu and Y.Watase, Phys. Rev. Lett. 33, 853 (1974).

24) M.Bernheim, A.Bussière, A.Gillebert, J.Mougey, Phan Xuan Hô, M. Priou, D.Royer, I.Sick and G.J.Wagner,Phys.Rev.Lett. 32,898 (1974)

25) J.Mougey, M.Bernheim, A.Bussière, A.Gillebert, Phan Xuan Hô, M. Priou, D.Royer, I.Sick and G.J.Wagner, preprint, Saclay 1975.

26) D.S.Koltun, Phys. Rev. Lett. 28, 182 (1972).

27) A.Faessler, S.Krewald and G.J.Wagner, Phys. Rev. C11,2069 (1975).

28) S.Boffi, C.Giusti, G.D.Pacati,preprint, Pavia 1976.

MANY-BODY THEORY OF THE OPTICAL-MODEL POTENTIAL[†]

J.-P. JEUKENNE, A. LEJEUNE and C. MAHAUX,

Institut de Physique, Université de Liège

4000 Liège I, Belgium

Abstract. We describe a calculation of the complex optical-model potential that is based on Brueckner's theory and on Reid's hard core interaction. We discuss successively nuclear matter and finite nuclei. The symmetry potential and the Coulomb correction due to the nonlocality are also computed and compared with empirical values.

1. Introduction

In an accompanying paper,we give a brief survey of several theoretical approaches to the optical-model potential (OMP) . There, we insist particularly on the interest of those calculations which are based on realistic nucleon-nucleon interactions. In the present paper, we sketch the basic ideas and results of (mainly) one such investigation, which uses as sole inputs Reid's hard core nucleon-nucleon interaction [1)] and the empirical density distribution in finite nuclei. We pay special attention to those aspects of our results which yield information that is not attainable from analysis of the experimental data. In Sect. 2, we describe the main lines of our theoretical approach. Then, we discuss successively the real part of the OMP in nuclear matter (Sect. 3) and in finite nuclei (Sect. 5), the imaginary part of the OMP in nuclear matter (Sect. 4) and in finite nuclei (Sect. 5), the role of the Coulomb interaction (Sect. 6) and the symmetry potential associated with neutron excess (Sect. 7).

---

† presented by C. Mahaux

2. Theoretical approach

2a. Definitions and properties [2)]

Let us first consider infinite, uncharged and symmetric nuclear matter. If the nucleons would only feel an average potential V_k ($V_k > 0$), which as indicated may depend on the nucleon momentum k , their energy would be ($\hbar = 1$)

$$e(k) = \frac{k^2}{2m} - V(k) \quad . \tag{1}$$

In this free Fermi gas model, the nucleons fill all momentum states up to the maximum value k_F , called the Fermi momentum. One has the relation

$$\rho = \frac{2}{3\pi^2} k_F^3 \quad , \tag{2}$$

where ρ is the matter density.

In reality, the nucleons interact with one another and the true (normalized) ground state $|0\rangle$ of nuclear matter is correlated. Let us create a nucleon with momentum k at time t = 0 on top of this correlated ground state : $a^\dagger(k,t = 0)|0\rangle$.
The probability amplitude that at a subsequent time t > 0 the created nucleon has not undergone any collision is given by the "one-body Green function"

$$G(k,t) = - i \langle 0|a(k,t)\, a^\dagger(k,t = 0)|0\rangle \quad , \tag{3}$$

where the factor (- i) is conventional. In the optical model, one assumes that this amplitude decays exponentially in time:

$$G(k,t) = - i\, R_k\, e^{-\frac{t}{2\tau}}\, e^{-i\, e(k)\, t} \quad , \tag{4}$$

where e(k) is the energy of the nucleon with momentum k.
One has

$$e(k) = \frac{k^2}{2m} - V_k \quad , \quad \tau = (2\, W_k)^{-1} \quad , \tag{5}$$

where $- V_k - i W_k$ is the OMP. A Fourier transform of Eq. (4) over the variable t yields

(6) $$G(k,E) = \frac{R_k}{E - e_k + i W_k} \quad .$$

Thus, the pole $e_k - i W_k$ of $G(k,E)$ gives the real and the imaginary parts of the OMP.

The "mass operator" $M(k,E)$ is defined by

(7) $$G(k,E) = \{E - \frac{k^2}{2m} - M(k,E)\}^{-1} \quad .$$

Equations (6) and (7) show that the complex quantity $e_k - i W_k$ is the root of the energy momentum relation

(8) $$E = \frac{k^2}{2m} + M(k,E) \quad .$$

Comparing now Eqs. (5) and (8), one sees that the complex quantity $M(k,e_k)$ can be identified with the OMP felt by a nucleon with momentum k, or equivalently (inasmuch as the optical model is valid) by a nucleon with energy e_k.

In fact, the operator $M(k,E)$ has a meaning even for $E \neq e_k$: let us perform a Fourier transform over the variable k . This yields the <u>nonlocal and energy-dependent</u> OMP $M(|\vec{x} - \vec{x}'|,E)$. Hence, the variable k is related to the "true" nonlocality and the variable E to the "true" energy dependence of the mean field.[3)]

The mass operator $M(k,E)$ is also a function of the density ρ : $M_\rho(k,E)$. By taking the experimental density distribution $\rho(r)$, one thus obtains M as a function of the distance r from the nuclear centre. This is the "local density approximation" that we shall use later on to construct the OMP in finite nuclei.

We summarize : The problem consists in calculating $M_\rho(k,E)$. Its Fourier transform over k yields the nonlocal, energy-dependent

OMP $M_\rho(|\vec{x} - \vec{x}'|,E)$. In most practical applications, one only makes use of the "on-shell" value $M(k,e_k)$, where the energy e_k and the momentum k are related by Eq. (8). If one considers $M(k,e_k)$ as a function of the energy e_k of the nucleon, one has a local, energy-dependent OMP, as in most empirical analysis : we then write $M(k,e_k) = M(e_k)$. If one considers $M(k,e_k)$ as a function of k, one gets by Fourier transformation a nonlocal, energy-independent OMP as used for instance by Perey and Buck:[4] we then write $M(k,e_k) = M(k)$.

For $k = k_F$, the root of Eq. (8) is called the "Fermi energy", ε_F. The following substracted dispersion relation holds

(9) $$V(k,E) = - f(k) + \frac{P}{\pi} \int_{-\infty}^{\infty} dE' \; \frac{W(k,E')}{E' - E} \quad ,$$

provided that a suitable definition of $M(k,E) = - V(k,E) - i\, W(k,E)$ is used for $E < \varepsilon_F$. One also has the asymptotic behaviour, for $E \to \varepsilon_F$

(10) $$W(k,E) \propto (E - \varepsilon_F)^2 \quad .$$

2b. Computational approaches

The problem thus amounts to calculate the mass operator $M_\rho(k,E)$ in nuclear matter; for simplicity, we usually omit the index ρ. Several computational procedures have been proposed :

(I) If one is satisfied with the use of a weak effective nucleon-nucleon interaction $\hat{v}$, perturbation theory can be employed. Its leading term is the "Hartree-Fock" approximation (HF)

(11) $$\hat{M}_{HF}(k) = \sum_{j<k_F} \langle jk|\hat{v}|jk - kj\rangle \quad .$$

where $|jk\rangle$ is a product of two plane waves and where spin and isospin quantum numbers are implicit. Note that the HF approximation (11) is static, i.e. independent of E (it yields, however, a nonlocal potential). Moreover, the HF potential (11) is <u>real</u>: one must go to higher-order terms to obtain an imaginary part; these higher-order terms also introduce a dependence on E, i.e. dynamical effects. This is not surprizing: Eq. (9) shows that the existence of an imaginary part is directly related to the property that the OMP depends on the "true" energy E. The trouble with the perturbation expansion is that $\hat{v}$ is an adjusted effective interaction, so that its use may blur the functional dependence of M(k,E) on k and E. Moreover, its meaningfulness is limited to rather low energy. Therefore, perturbation theory was used mainly in early calculations (see, e.g., Refs. 5 to 8).

(2) At high energy, one can use the "impulse approximation". In the case of a central nucleon-nucleon force, this approximation reads

$$(12)\qquad M_{IA}(e) = \pi(4\ i\ m\ k)^{-1}\ \rho \sum_{T,L,J} (2T+1)\ (2\ J+1) \left(\exp 2\ i\ \delta(LJT) - 1\right) \quad ,$$

where L, J and T refer to orbital and total angular momentum and to total isospin, respectively, while $e = k^2/2m$ and δ is the nucleon-nucleon phase shift. The IA neglects binding and off-shell effects and the Pauli principle; it is therefore reliable only at high energy (see Fig. 8 below). One can, however, include the role of the Pauli principle in an approximate way, so that for instance the low energy behaviour (10) is (approximately) reproduced: see the pioneering work by Clementel and Villi.[9] Other improvements can bring the domain of validity of the (thereby modified) impulse approximation down to about 100 MeV. The IA has mainly been used for light targets.

(3) In the case of a realistic (i.e. strong) nucleon-nucleon interaction, perturbation theory is not valid. All approximation schemes

which have been used to date (for the calculation of the OMP) try to make use of the short-range nature of the nuclear force, which has the consequence that three nucleons rarely interact simultaneously:

(a) Some calculations are based on the hierarchy of equations which couple on one-, two-, ... -body Green functions;[10] this infinite set of equations must in practice be truncated. This can be made in a variety of ways and yields the socalled Λ_{00}, Λ_{10}, Λ_{11} approximations. For a review, see Ref. 11; for recent applications, see Refs. 12 to 14. It appears that the Λ_{00} and Λ_{10} approximations are not reliable at low energy, at least as far as the imaginary part of the OMP is concerned; the Λ_{11} approximation is complicated and has not been applied yet.

(b) Most nuclear calculations based on realistic interactions involve the so-called "Brueckner" theory, which essentially consists in rearranging the perturbation series by grouping terms of the same order in the density ρ. This type of approach has met with considerable success for a large number of problems (see Refs. 15 to 17). It is therefore the one that we have adopted in a series of papers [2,18-22] where details can be found. Below, we only sketch some of these previous results, and we show a number of new ones. The leading term of the expansion is usually called the Brueckner-Hartree-Fock (BHF) approximation. For technical reasons, we denote it by $M_1(k,E)$:

$$M_{BHF}(k,E) = M_1(k,E) = \sum_{j<k_F} \langle jk|g(E+e_j)|jk - kj\rangle \quad , \tag{13}$$

where (see Eqs. (1) and (8))

$$e_j = \frac{j^2}{2m} + \mathrm{Re}\, M_1(k,e_j) \quad . \tag{14}$$

The operator g(w) in Eq. (13) is Brueckner's reaction matrix; it is the solution of a (so-called Bethe-Goldstone) integral equation, suitably generalized to the case of scattering:[18] it is then a complex operator. Note that the BHF approximation (13) yields an OMP which is nonlocal <u>and</u> energy-dependent. All results shown below have been calculated from Reid's hard core nucleon-nucleon interaction.[1]

3. Real part of the OMP in nuclear matter

We calculated

$$V(e_k) = - \text{Re}\, M_1(k,e_k) \tag{15}$$

for various values of the density, i.e. of k_F (see Eq. (2)). The energy dependence of V is shown in Fig. 1 (where $E = e_k$), for

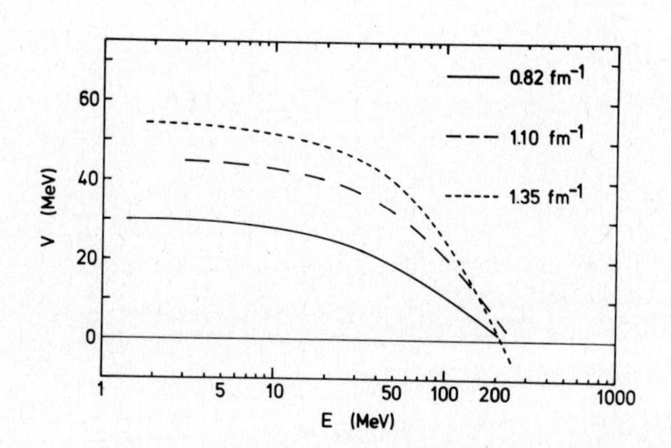

Fig. 1. Dependence on the bombarding energy $E=e_k$ (see Eq.(1)) of the calculated depth of the OMP in symmetric nuclear matter, for the Fermi momenta 1.35, 1.10 and 0.82 fm^{-1}, respectively.

three values of k_F, which correspond to $\rho = 0.166\ fm^{-3}$ ($k_F = 1.35\ fm^{-1}$ central density), $\rho = 0.090\ fm^{-3}$ ($k_F = 1.10\ fm^{-1}$) and to $\rho = 0.037\ fm^{-3}$ ($k_F = 0.82\ fm^{-1}$), respectively. We see that the potential depth decreases with decreasing ρ, as expected. It is, however, not strictly proportional to ρ : this corresponds to the fact that the "effective" interaction is density-dependent. Relatedly, it leads to the observation that the half-potential radius is larger than the half-density radius (see Sect. 5). In Fig. 2, we compare our calculated results at

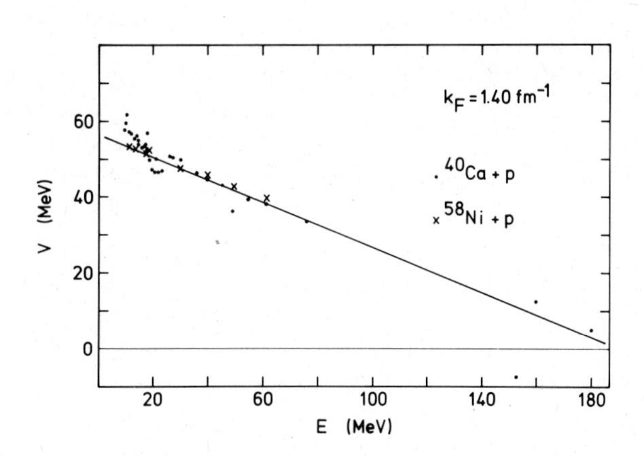

Fig. 2. Comparison between the calculated depth of the OMP in nuclear matter (full lines) and empirical values in the case of p + ^{40}Ca (Ref. 23) and p + ^{58}Ni (Ref.24) .

$k_F = 1.40\ fm^{-1}$ ($\rho = 0.185\ fm^{-3}$) to empirical OMP depths, taken from Refs. 23 and 24.

As noted in Sect. 2, the Fourier transform of $V_k = - U_k$ is a nonlocal potential. This Fourier transform is represented by the dashes in Fig. 3 ($s = |\vec{x} - \vec{x}'|$) where the full curve is a Gaussian with a nonlo-

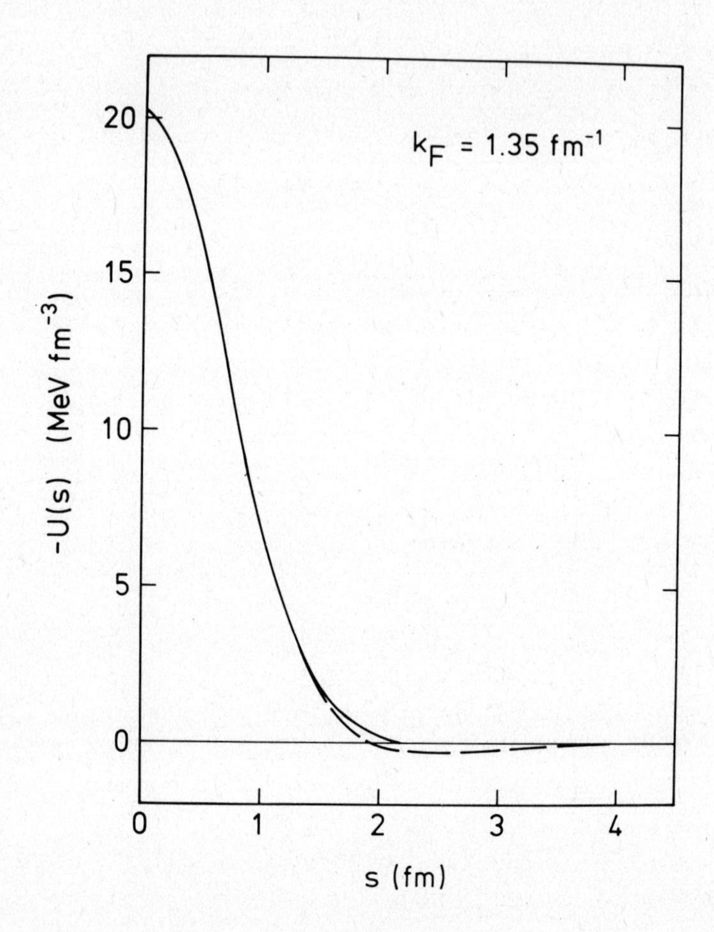

Fig. 3. The dashes show the dependence on $s=|\vec{r}-\vec{r}'|$ of the Fourier transform of our calculated depth $V_k = -U_k$ of the OMP; the full line is a Gaussian fit to the dashed curve, with nonlocality 1.0 fm .

cality range 1.0 fm . The good agreement corroborates the phenomenological model of Perey and Buck;[4] the small discrepancy indicates, however, that the Gaussian assumption is not fully justified. In other words, the nonlocality is slightly energy-dependent : it is found equal to 0.84 fm at e = 7 MeV, and to 0.70 fm at e = 50 MeV, for k_F = 1.35 fm^{-1}. Note that this nonlocality differs from the true nonlocality because it includes a contribution from the true energy dependence of M(k,E); the true nonlocality range equals 1.02 fm at 7 MeV.[2]

4. Imaginary part of the OMP in nuclear matter

In Fig. 4, we show the dependence on energy ($E = e_k$) of

$$W(e_k) = - \text{Im } M_1(k,e_k) \quad , \tag{16}$$

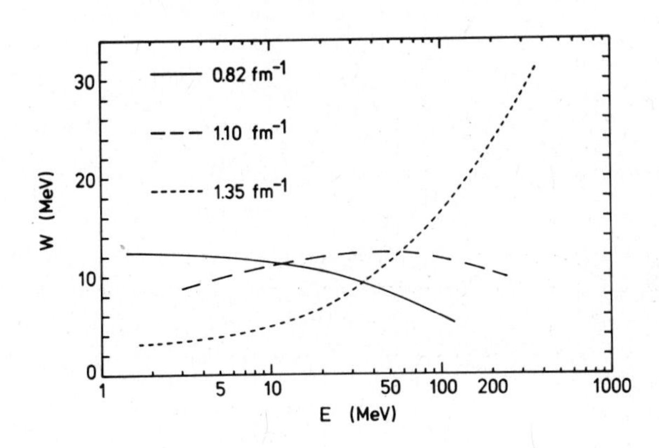

Fig. 4. Same as Fig. 1, but for the imaginary part of the OMP.

for three values of k_F . We see that below 50 MeV the absorptive part of the OMP is largest at the nuclear surface, while the opposite holds true at higher energy. This is in keeping with empirical evidence. One can study the true nonlocality of Im $M_1(k,E)$, but this does not offer much interest because the observed energy dependence of the imaginary part of the local OMP is dominated by the true energy dependence of the OMP. Note that this is at variance with beliefs based on empirical analysis.[4)]

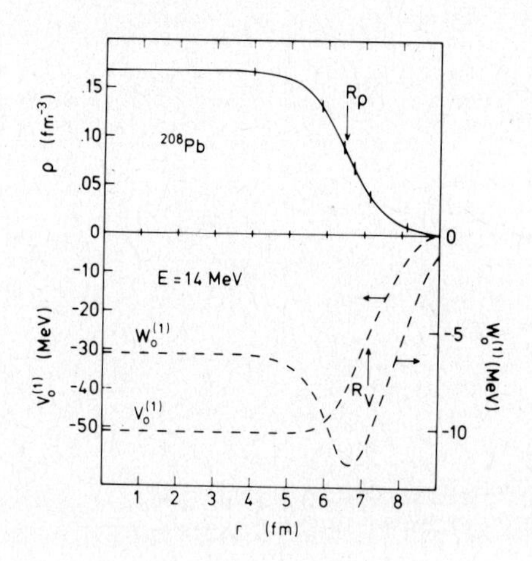

Fig. 5. The upper part of the figure shows the density distribution in ^{208}Pb, taken from Ref. 25. The lower part shows the real (left-hand scale) and imaginary (right-hand scale) parts of the OMP, as calculated from the Brueckner-Hartree-Fock plus local density approximations. The arrows R_ρ and R_V point to the half-density and half-potential radii, respectively.

5. OMP in finite nuclei

We now construct the OMP in a finite nucleus from the local density approximation (LDA) sketched in Sect. 2. The upper part of Fig. 5 represents the density distribution $\rho(r)$ in ^{208}Pb, taken from Ref. 25. The arrow R_ρ shows the half-density radius. The lower part of Fig. 5 shows the radial dependence of the real ($V_o^{(1)} \equiv - V_1$) and imaginary ($W_o^{(1)} \equiv - W_1$) parts of the OMP at 14 MeV. Note that the half-potential radius R_V is 0.65 fm larger than the half-density radius.

The theoretical curves should still be modified to include the Coulomb interaction (in the case of protons) and the symmetry potential

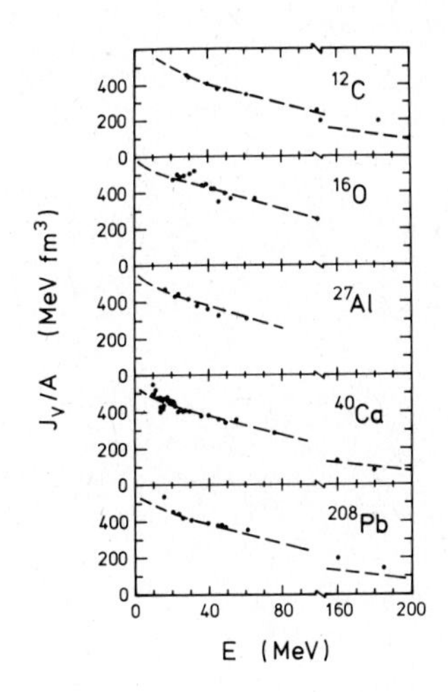

Fig.6. The long dashes represent the calculated volume integral per nucleon of the real part of the OMP, including the contributions of the symmetry and Coulomb components (preliminary results). The full dots are empirical values.[26)]

(which arises from neutron excess). These corrections are discussed below and are included in Fig. 6, where we show our theoretical results for the volume integral per nucleon (long dashes) of the real part of the OMP, together with empirical values (full dots) taken from Ref.26. We see that the agreement is quite satisfactory. Similar agreement is obtained between theoretical and empirical root mean square radii.[20]

In Fig. 7, we compare our theoretical results for the volume integral per nucleon of the imaginary part of the OMP with empirical values compiled in Ref. 27 for mass numbers larger than 40. Here again the agreement is quite satisfactory.

Finally, we show in Fig. 8 the difference between our results (full lines) based on the Brueckner-Hartree-Fock approximation (13) and the results obtained from the impulse approximation (12) (dashes). We see that the difference is mainly spectacular for the imaginary part of

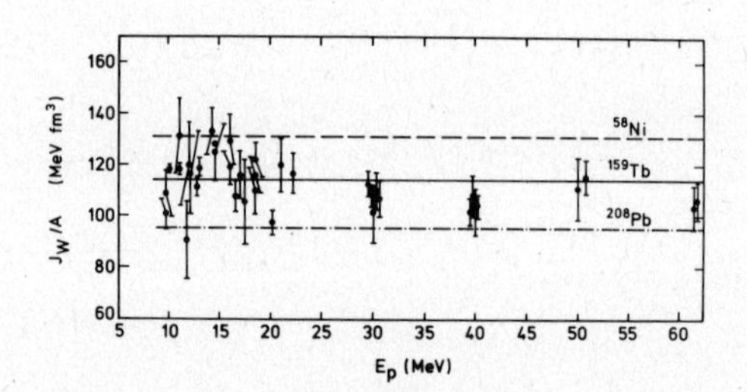

Fig. 7 - Comparison between the calculated volume integral of the imaginary part of the OMP and a compilation of empirical values.[27] (preliminary).

the OMP. It can be shown that the two approximations should become equal in the high-energy limit.[18] However, Fig. 8 shows that there still exists some difference at 300 MeV; it is mainly due to the binding correction, wich corresponds to the fact that the target nucleons (j in Eq. (13)) are bound.

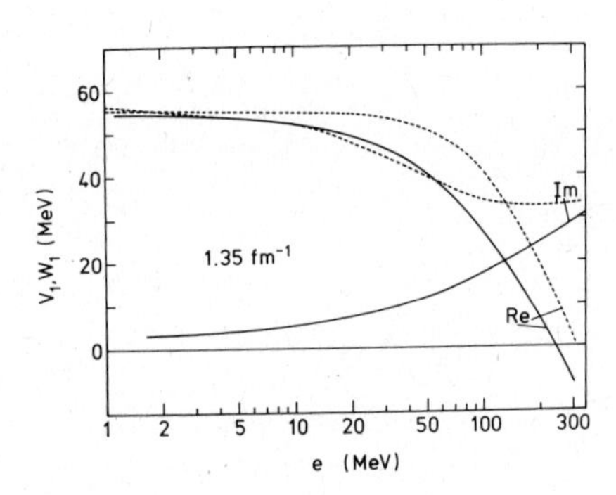

Fig. 8. Comparison between the Brueckner-Hartree-Fock approximation (13) (full lines) and the impulse approximation (12) (dashes), for k_F=1.35 fm^{-1}.

6. Coulomb correction

In first approximation, the role of the Coulomb field is simply to add to the real part of the OMP $M(k,E-V_C)$ the central potential V_C created by a uniformly charged sphere. Without going into details, the argument runs as follows. Let us first switch off the Coulomb interaction; the OMP $M(k,E)$ is nonlocal and energy-dependent; the corresponding empirical OMP is obtained by taking $M(k,E)$ at the value $E = e_k$ which is a root of (8) (we only take the real part, for simplicity). If a Coulomb potential V_C is added to the right-hand side of (8), the relation between E and k is modified. One can show that the resulting real OMP is given by Re $M(k,e_k) + V_C - \Delta_C$, where the Coulomb correction Δ_C arises from the total energy dependence of the OMP.

In Fig. 9, we compare the standard Coulomb correction (dashes)which was taken equal to $0.4\ Z/A^{1/3}$ MeV at the nuclear centre, to the one that we calculated (at 25 MeV) from the total energy dependence of our OMP.[29)]

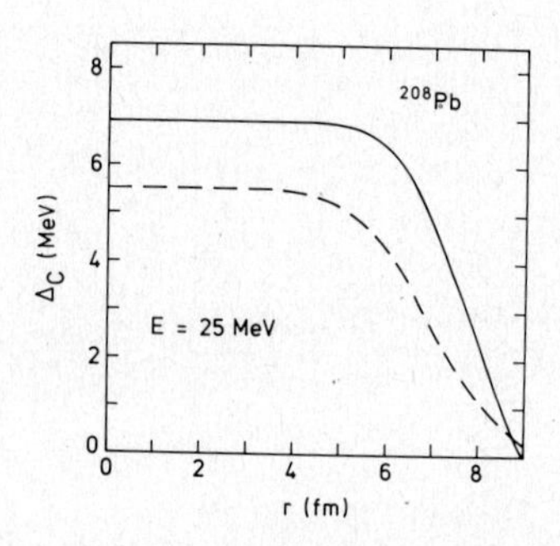

Fig.9. Calculated (full line) and standard (dashes) Coulomb corrections, in the case of ^{208}Pb at 25 MeV.

7. Symmetry potential

In the presence of neutron excess, the OMP is different for protons and for neutrons, respectively, even if the Coulomb field is disregarded. One writes thus the OMP for nucleon of type N (N = n or p) in the form

$$(17) \qquad V^{(N)} = V^{(o)} \mp U_1^{(N)} \alpha + (V_C - \Delta_C) \quad \delta_{Np} \quad ,$$

where $\alpha = (N - Z)/A$ is the asymmetry parameter; $(V_C - \Delta_C)$ is the corrected Coulomb field (see Sect. 6) and the upper sign refers to protons. The radial dependence of all quantities is implicit. The symmetry potential U_1 can be obtained from the analysis of proton scattering by a wide range of nuclei: this yields $U_1^{(p)} \simeq 24$ MeV if the standard value $(0.4\ Z/A^{1/3})$ is used for Δ_C.[28] If, however, the more correct value (see Fig. 9) is used for the Coulomb correction Δ_C, we obtain $U_1^{(p)} \simeq 12.3$ MeV.The latter value is in fair agreement with the value of $U_1^{(n)}$ ($\simeq 12.5$ MeV) obtained from neutron scattering,[30] while a worrying

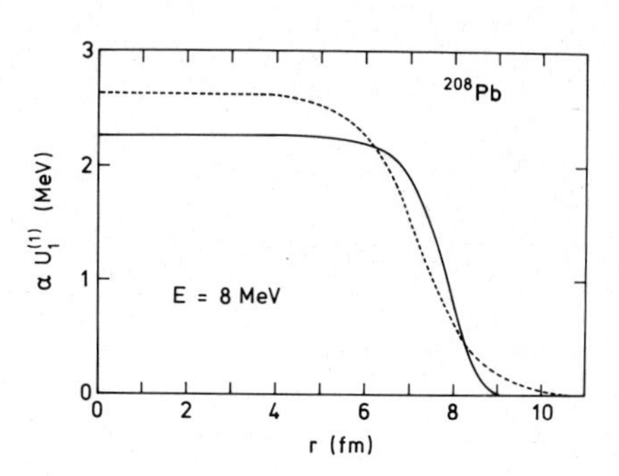

Fig.10. Comparison between the calculated symmetry part of the OMP for 8 MeV neutrons on ^{208}Pb (full curve) and the empirical value (dashed curve), taken from Ref. 30.

disagreement was clearly found if the standard value of Δ_C was used.

We have calculated the value of U_1 from the BHF approximation.[29] Our results are plotted in Fig. 10 (full curve), in the case of ^{208}Pb at 8 MeV. The long dashes represent the empirical symmetry part of the OMP, taken from Ref. 30. We see that the agreement is quite good: the theoretical volume integral per nucleon of the symmetry part U_1 α of the potential is 21.8 MeV fm^3, while the empirical value is 21.9 MeV fm^3. Finally, we show in Fig. 11 the calculated value of the sum α U_1 + Δ_C(full curve) and the empirical one,[31] in the case of ^{208}Pb at 25 MeV.

8. Conclusion

We have seen in Sects. 3-5 that the Brueckner-Hartree-Fock (BHF) approximation is remarkably successful in reproducing the observed

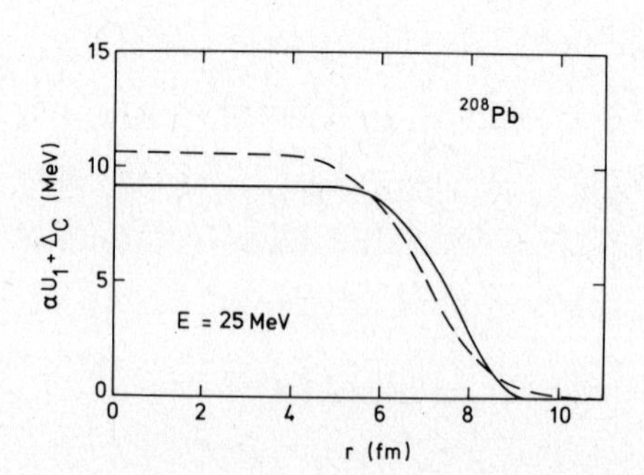

Fig.11. Comparison between the calculated (full curve) value of the sum α U_1 + Δ_C(see eq.(17)) and the empirical one,[31] in the case of ^{208}Pb + p at 25 MeV.

features of the empirical optical-model potential. We recall that no parameter was adjusted in our calculation. This has led us to investigate fine details of the OMP, such as the Coulomb correction and the symmetry term (Sects. 6,7).

Criticisms have recently been raised against the use of the BHF approximation for the calculation of the binding energy of nuclear matter, see e.g. Ref. 32. However, these criticisms only question the accuracy of the BHF for $k_F > 1.3\ fm^{-1}$, while our results concern the region $k_F < 1.35\ fm^{-1}$. Hence, we do not believe that these recent works, even if substantiated by forthcoming investigations, would impair the significance of our results. Moreover, the calculation of the binding energy per nucleon in the frame of the BHF approximation is essentially based on the assumption that the independent particle model is correct.[33] As recently emphasized [2,33] this is not accurate. However, the calculation of the OMP precisely deals with only that fraction (= 70 %) of the full wave function which is described by the independent particle model: one does not have to assume that this fraction is very large. In other words, it may happen that BHF approximation can be used for the calculation of the OMP while being inaccurate in the case of the binding energy.

References

1) R.V.Reid, Ann.Phys. (N.Y.) 50, 411 (1968).

2) J.-P.Jeukenne, A.Lejeune and C.Mahaux, Physics Reports (1976).

3) M.Bertero and G.Passatore, Z.Naturforsch. 28a, 519 (1973).

4) F.Perey and B.Buck, Nucl.Phys. 32, 353 (1962).

5) L.Verlet and J.Gavoret, Nuovo Cim. 10, 505 (1958).

6) B.Jancovici, Nucl. Phys. 21, 256 (1960).

7) B.Jancovici, Prog.Theor.Phys. (Kyoto) 23, 76 (1960)

8) G.Ripka, Nucl. Phys. 42, 75 (1963).

9) E.Clementel and C.Villi, Nuovo Cim. 2, 176 (1955).

10) P.C.Martin and J.Schwinger, Phys. Rev. 115, 1342 (1959).

11) M.Weigel and G.Wegmann, Fortschritte der Physik 19, 451 (1971).

12) Q. Ho-Kim and F.C.Khanna, Ann.Phys. (N.Y.) 86, 233 (1974).

13) H.Gall and M.R.Weigel, Z.Physik A276, 45 (1976).

14) C.Marville, preprint (Liège, 1976).

15) H.A.Bethe, Annu.Rev.Nucl.Sci. 21, 93 (1971).

16) D.W.L.Sprung, Advances in Nuclear Physics, edited by M. Baranger and E.Vogt (Plenum Press, N.Y., 1972) 5, 225 (1972).

17) H.S.Köhler, Phys. Reports 18, 217 (1975).

18) J.Hüfner and C.Mahaux, Ann.Phys. (N.Y.) 73, 525 (1972).

19) J.-P.Jeukenne, A.Lejeune and C.Mahaux, Phys.Rev. C10, 1391 (1974).

20) J.-P.Jeukenne, A.Lejeune and C.Mahaux, in Proceedings of the International Conference on Nuclear Self-Consistent Fields (Trieste, February 1975), Edited by G.Ripka and M.Porneuf, p.155 (North-Holland Publ. Comp., Amsterdam, 1975).

21) J.-P.Jeukenne, A.Lejeune and C.Mahaux, Nukleonika 20, 181 (1975).

22) J.-P.Jeukenne, A.Lejeune and C.Mahaux, Phys. Letters 59B, 208 (1975).

23) W.T.H. Van Oers, Phys. Rev. C3, 1550 (1971).

24) G.L.Thomas and E.J.Burge, Nucl.Phys. A128, 545 (1969).

25) M.A.Preston, Physics of the Nucleus (Addison-Wesley Publ. Comp. Inc., Reading, Mass., 1962).

26) W.T.H. Van Oers, H.Haw, N.E.Davison, A.Ingemarsson, B.Fagerström and G.Tibell, Phys. Rev. C10, 307 (1974).

27) D.C.Agrawal and P.C.Sood, Phys. Rev. 9C, 2454 (1974).

28) G.R.Satchler, in Isospin in Nuclear Physics (North-Holland Publ. Comp., Amsterdam, 1969) ch. 9

29) J.-P.Jeukenne, A.Lejeune and C.Mahaux, to be published.

30) B.Holmqvist and T.Wiedling, Nucl. Phys. A188, 24 (1972).

31) F.D.Becchetti and G.W.Greenlees, Phys. Rev. 182,1190 (1969).

32) J.C.Owen, R.F.Bishop and J.M.Irvine, Phys. Letters 59B, 1 (1975).

33) R.S.Poggioli and A.D. Jackson, Phys. Rev. Letters 35, 1271 (1975).

OPTICAL MODEL POTENTIAL AND NUCLEAR DENSITY DISTRIBUTIONS

P.E. HODGSON

Nuclear Physics Laboratory, Oxford

Abstract. Folding model calculations of the optical model potentials of alpha-particles and heavy ions are reviewed, showing the importance of accurate density distributions. The method of calculating the density distributions from single-particle potentials is described, and the contributions of some smaller effects discussed.

1. Introduction

One of the most important lessons of the vast effort over the last twenty years devoted to fitting elastic scattering cross sections with optical potentials is that the more physics one builds into the potential from the beginning the more acceptable the resulting potential is likely to be. It is often quite possible to fit the data with bizarre potentials that no-one would accept as physical, so fitting is usually carried out with a Saxon-Woods form for the real part of the potential, and this satisfies our simple intuition that the potential should be uniform in the nuclear interior and fall exponentially to zero around the nuclear surface, because the nucleon-nucleon force is short-range and falls to zero in this way. Although it is not certain that such a form is always adequate in the surface region it has been widely used with considerable success. The absorbing part of the potential is less well understood, and a variety of forms have been used ranging from volume to surface-peaked, and the goodness of fit obtained probably owes more to the general insensitivity of the cross-sections to the precise form of the absorbing potential than to its physical realism.

The resulting potentials contain parameters that should be connected with our knowledge of nuclear structure and of the nucleon-nucleon interaction, and in so doing we may hope to improve our knowledge of the optimum form.

This has been done in a series of calculations of the optical potential from the nuclear density distribution $\rho(r)$ and the nucleon-nucleon interaction $v(r)$. For nucleon-nucleus scattering this takes the simple form

$$(1.1) \qquad V(r) = \int \rho(\underline{r}')\, v\,(|\underline{r} - \underline{r}'|)\,d\underline{r}' \quad ,$$

while for nucleus-nucleus scattering one can either use a single folding with the nucleon-projectile potential for $v(r)$, or a double folding with the nucleon-nucleon interaction

$$(1.2) \qquad V(r) = \iint \rho_1(\underline{r}_1)\rho_2(\underline{r}_2)\, v\,(|\underline{r}+\underline{r}_1-\underline{r}_2|)\,d\underline{r}_1 d\underline{r}_2 \quad .$$

This double folding has the advantage of treating both nuclei symmetrically.

In the earlier studies by Greenlees, Pyle and Tang[1] various forms of the nucleon-nucleon interaction were used in (1.1) together with reasonable density distributions to calculate the differential cross-section for elastic scattering.

This simple calculation cannot be expected to give precise results, and it may be improved either by formulating and evaluating a series of correction terms to (1.1) and (1.2), or by regarding v as an effective nucleon-nucleon interaction and evaluating it by the techniques of many-body theory. Both approaches have been used, and some of the work along these lines is reviewed in Sect. 2. Attention is concentrated on potentials for alpha-particles and heavy ions because folding models are likely to be more reliable in the surface region and the scattering of such particles is especially sensitive to the potential there.

As the accuracy of this work improves, it becomes necessary to use the best available nuclear density distributions. These may be obtaine in a wide variety of ways[2] and, as for the optical potentials, the parametrised forms fitted accurately to particular experimental data are not always the most physically acceptable. For example, there was at one time much discussion about the form of the charge distribution in the centre of the nucleus, and many analyses were made using various parametrised forms with humps and dips in that region. It was subsequently realised that the optimum form was chosen mainly by the small improvements it effected in the knee region of the charge distribution which mainly determines the elastic scattering cross section, and that it gave little information about the central region.

These difficulties have to a large extent been overcome by the development of model independent ways of obtaining the charge distribution from the scattering cross-section.[3-7] This work has confirmed the previous result that the electron scattering and muonic atom data give charge distributions that are accurate in the knee region but are less well known in the centre and surface regions. This is shown for example by the experimental charge distribution of Fig. 1.1.

Such work has established that it is no longer adequate to use a Saxon-Woods form factor for the charge distribution; for many nuclei there is certainly a radial oscillation of the charge density in the nuclear interior, reflecting the shell structure of the nucleus.[8] This radial oscillation is only found if electron scattering measurements are made for momentum trasfers $q > 2.1$ fm^{-1} (Ref. 9). In general, the amplitudes of the Fourier components of $\rho(r)$ having wavelength less than $2\pi/q_{max}$ are not determined by experiment.

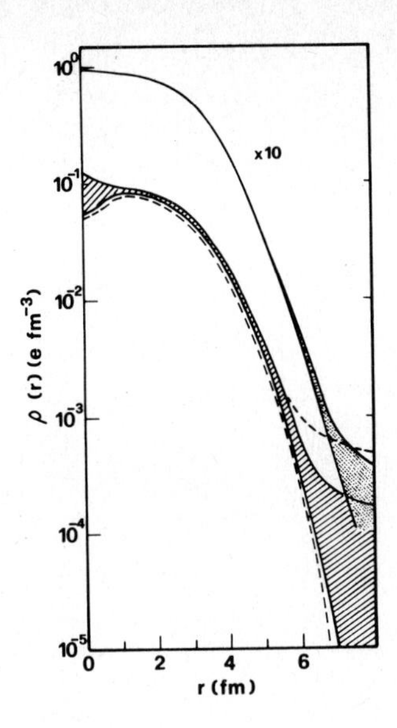

Fig.1.1. Charge distribution of ^{32}S obtained by analyzing electron ela stic scattering and muonic atom data and showing the accuracy of the determination at various radial distance. The two curves refer to analyses of data obtained at Mainz and Stanford, and the dashed curves for the former refer to limits obtained from the elastic scattering data above.The points on the Stanford distribution refer to the model density.[7)]

Nuclear matter distributions are less well known than the charge distributions, essentially because we know less about the nuclear force than about the Coulomb interaction.

These uncertainties make it desirable to explore alternative and more physical ways of determining the nuclear density distributions, and we consider in Sects. 3 and 4 those obtained by summing nucleon single particle (SP) wave functions calculated as eigenvalues of one-body potential.

Nuclear density distributions have also been obtained from many different types of Hartree-Fock calculations. This is a more fundamental theory, and uses less phenomenological information than the SP calculations, so it is possible that it gives better information on some features of the distributions, but they differ too widely among themselves

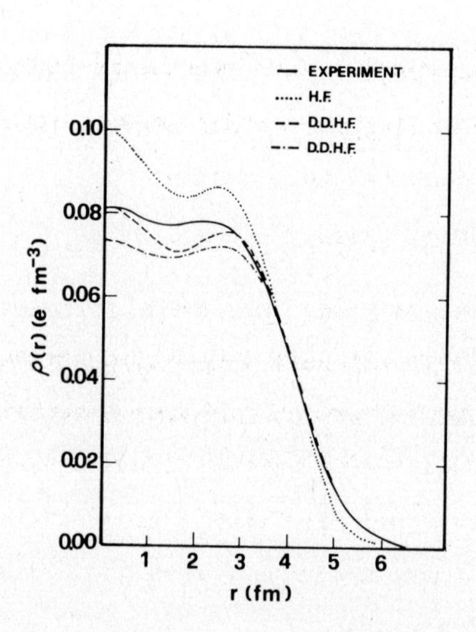

Fig. 1.2. Charge distribution of ^{58}Ni obtained from electron elastic scattering and compared with various Hartree-Fock calculations.[8]

to be an acceptable source of precise information (see Fig. 1.2). It is possible that they could be made so by the application of the same phenomenological constraints as are applied here to the SP distributions.

2. Folding Model Analyses of Elastic Scattering Alpha-particles and Heavy Ions

There have been many analyses of the elastic scattering of alpha-particles and heavy ions by nuclei, and these have been generally successful, though in many cases it has proved necessary to adjust some parameter of the folded potential to optimize the fit to the experimental data.

The simplest approximation is to assume that the nucleon-nucleon potential has zero range, and then the double-folding potential reduces to the overlap of the density distributions

$$V(r) = -\frac{2\pi h^2}{M} \bar{f} \int \rho_1(\underline{r}_1)\rho_2(|\underline{r}-\underline{r}_1|)d\underline{r}_1 \ . \qquad (2.1)$$

Vary and Dover [10] have used this expression to calculate the heavy ion potentials for a number of nuclei, using proton densities fitted to electron elastic scattering, and treating $\bar{f}$ as a complex adjustable parameter. Some typical fits to the data are shown in Fig. 2.1. The same potentials give good fits to the cross-sections for some one-nucleon transfer reactions, and rather less good fits to those of some two-nucleon transfer reactions.

More detailed calculations may be made with a phenomenological expression for the nucleon-nucleon interaction v(r). This is an effective interaction, as it refers to nucleons embedded in nuclei, and when (2.1) is used to calculate nucleon optical potentials it is appropriate to use an interaction obtained by solving the Bethe-Goldstone equa-

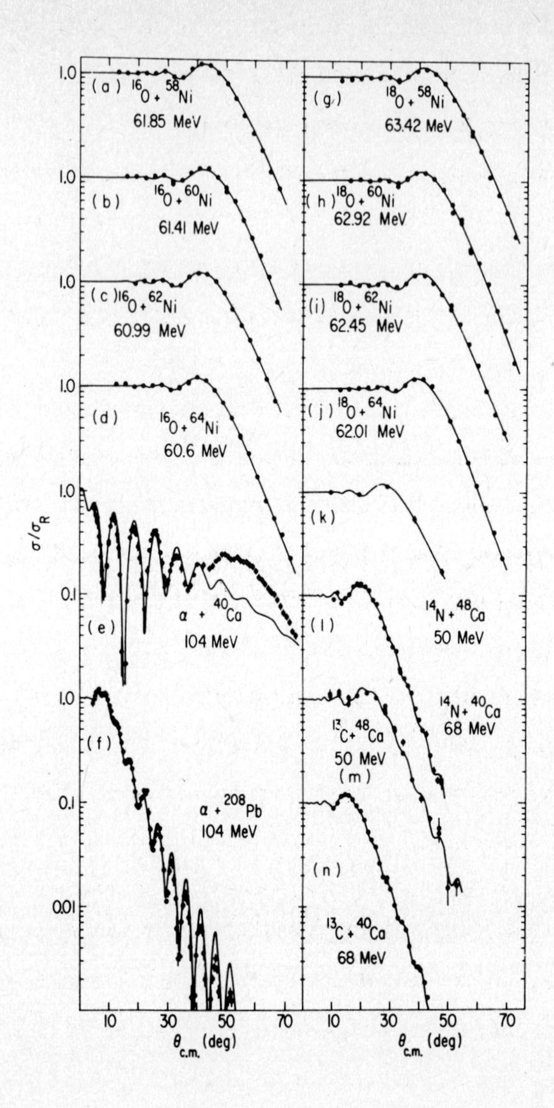

Fig.2.1. Alpha-particle and heavy ion elastic scattering cross-sections compared with calculations using a simple folding potential with adjustable complex strength.[11)]

tion. In the case of heavy ions, however, the main contribution to the integral in (2.1) comes from the nuclear surface, so the free nucleon-nucleon interaction may be more appropriate.

The double folding procedure has been used with success to analyse the elastic scattering of alpha-particles by nuclei, using standard nucleon-nucleon interactions and phenomenological imaginary potentials.[12-14] When applied to heavy ion scattering, however, it is found that the doubly-folded potential has to be normalized by a factor around 0.5 to fit the experimental data. Calculations by Satchler[15] also showed that for the interaction of ^{12}C and ^{208}Pb at 116.4 MeV (turning point 12.25 fm) the potential at distances less than 10 fm has no effect on the scattering for $\sigma/\sigma_R > 0.01$. Thus the interior potential has little effect on the scattering so that the folding potentials are of interest mainly in the exterior region.

A variety of folded potentials normalised to fit the same interaction are shown in Fig. 2.2 and it is notable that they all have a depth of 2.2 MeV at 11.85 fm. The Saxon-Woods potentials, fitted to the same data show the same behaviour, except that the point of interaction is displaced; the reason for this difference is not understood.

The double folding model has been used by Eisen[16] to analyse the elastic scattering of ^{16}O by ^{48}Ca, ^{44}Ca, ^{42}Ca and ^{40}Ca at energies near the Coulomb barrier. Assuming a charge distribution of ^{40}Ca obtained from analyses of electron scattering, this gave the nuclear matter distributions of the other calcium isotopes. These are determined with the greatest sensitivity in the far surface region where the density is around a tenth of the central density, and Eisen found that ^{42}Ca and ^{44}Ca have very similar densities, but that there is a marked difference between those of ^{40}Ca and ^{42}Ca. That of ^{48}Ca is greatest, but falls

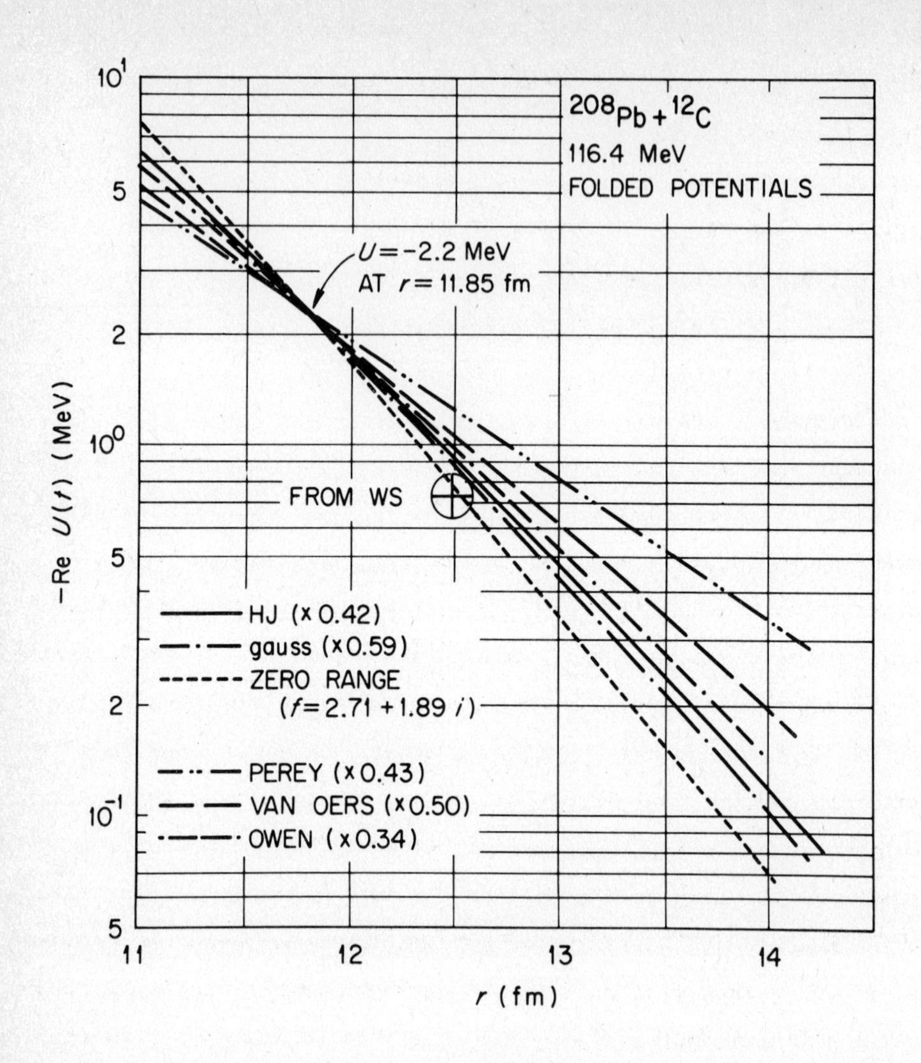

Fig.2.2.A series of double-folded heavy ion potentials fitted to the differential cross-section for the elastic scattering of ^{12}C on ^{208}Pb at 116.4 MeV showing their intersection at U= -2.2 MeV and r = 11.85 fm. The potentials were normalised by the factors in the brackets.[15)]

most rapidly with increasing radius. This analysis shows how heavy ion scattering is able to give detailed information on nuclear densities in the surface region.

There are several higher-order corrections to the simple folding expression that may be evaluated to give more realistic potentials:

a) Energy dependence of the nucleon-nucleon potential, and the motion of the nucleons in each nucleus.

b) The Pauli principle.

c) Three-body forces.

These corrections have been evaluated for light projectiles but so far there have been few calculations for heavy ions. As they certainly affect the calculation of heavy ion potentials it is appropriate to review them here:

a) Phenomenological optical potentials depend on the incident energy, and this is largely a consequence of the non-locality of the interaction. It is therefore usual to evaluate the nucleon-nucleon interaction at the incident energy divided by the number of particles in the projectile. This procedure ignores the effect of the motion of the nucleons in the target: since it is the momenta that add vectorially and the nucleon-nucleon interaction depends linearly on the energy, the effective energy of each nucleon-nucleon interaction is increased by that of the target nucleon, and this must be averaged over the target nucleus. Perkin et al.[17)] have evaluated this effect, and find that it reduces the strength of the potential by 3% for deuterons, 9% for helions, 10.6% for alpha-particles and by 5.2% for ^{12}C ions.

b) The Pauli principle is largely responsible for the reduction of the potential depth at low energies. When the incident particle enters the nucleus it can only occupy vacant states. At low energies many of these states are full, so the particle has difficulty in entering, and this is represented by increasing the repulsion of the potential, i.e. by making it shallower.

At higher energies more states are available so the necessary reduction is not so large. The energy of the particle inside the nucleus itself depends on the incident energy and on the potential, so the calculation must be made in a self-consistent way. Perkin et al.[17] obtain a maximum reduction of 11% for deuterons, wich is consistent with the estimate of 10-20% obtained by Perey and Satchler.[18]

The Pauli principle has been used by Block and Malik[19] to obtain a

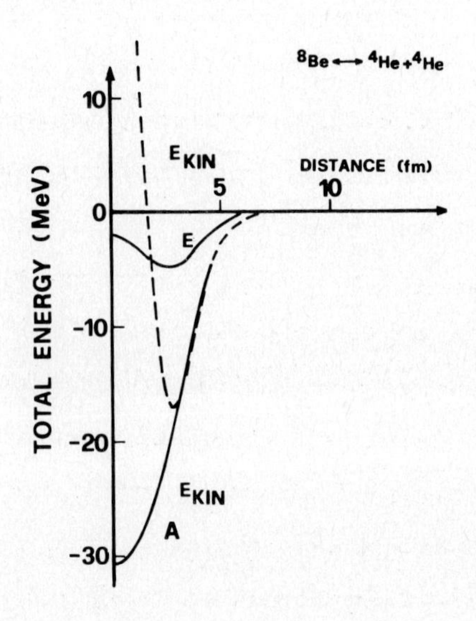

Fig. 2.3. Total energies of the α-α system showing the effects of antisymmetrising the total wave function and of subtracting the intrinsic kinetic energy.[21]

shallow potential for the $^{16}O-^{16}O$ interaction, which they use to explain qualitatively some of the observed regularities in the excitation function.[20]

Calculation by Zint and Mosel[21] show that the reduction in the depth of the potential found in folding model calculations is largely due to the change in the kinetic energies of the nucleons due to the Pauli principle. Their results for the α-α potential are shown in Fig. 2.3: the curve labelled E is the expectation value of a Hamiltonian consisting of a kinetic energy term and a Skyrme potential, and that labelled $E-E_{kin}$ is obtained by subtracting from it the kinetic energy so that it corresponds to a folding model taking account of the Pauli principle. The dashed curve shows the result of a calculation without antisymmetrisation, i.e. without distortion of the density or kinetic energy contributions.
This agrees down to $R \approx 3.5$ fm with the corresponding result with antisymmetrisation, but becomes repulsive at small distances due to the incompressibility of nuclear matter.

Rook has also suggested that the effect of the Pauli Exclusion Principle can be calculated by including a factor exp $(i\underline{k}\underline{r})$ in the folding integral to take account of the relative motion of the interacting ions. In this way the number of nucleons that must be promoted to unoccupied states to satisfy Pauli principle can be determined and hence the effective repulsive potential at zero separation. The effects of this term has been studied by Perez[22] by comparing with the data on $^6Li-^6Li$ scattering from 9 to 16 MeV, but its inclusion did not give significantly better fits.

c) The three-body forces in a heavy ion interaction are of two types, depending on whether a pair of particles is in the target or in the projectile. In the former case they are included in the phenomenological nucleon-nucleus potential, and so do not have to be taken into account explicitly if the single folding expressions is used. Perkin et al.[17] used the local density approximation and reasonable forms for the t-matrix and the two-body correlation function, and obtain the results given in Table 2.1.

Table 2.1. Percentage reduction in the potential depth due to the energy dependence of the nucleon potential and to the three-body forces.[17]

Incident Particle	Energy Dependence	Three-body forces	Total
Deuteron	0.3	0.1	3.1
Helion	9.0	3.6	12.6
Alpha-particle	10.6	3.6	14.2
^{12}C	5.2	0.8	6.0

The folding model calculations of heavy ion reactions tend to give values of the surface diffuseness parameters that are too high. Rook and Perkin have investigated this effect and find that it can be understood as a result of the excitation of the target nucleus by the incoming particle.

Several calculations with the folding model have been made by Rowley[23] and using a density distribution of Saxon-Woods form with $R_1=1.04 \times A^{1/3}$, $a_1=0.54$ he finds an optical potential with $V=50$ MeV, $R_2=1.15 \times A^{1/3}$, $a_2=0.65$ which fits quite well a number of elastic scattering cross-sections from that of ^{16}O on iron to that of krypton and thorium. This model is quite good near the Coulomb barrier which is the most important region but gives potentials that are far too deep in the centre.

The extent of the region of absorption depends on the strength of the imaginary part of the interaction potential, and this may be studied by the Coulomb-nuclear interference effects in anelastic scattering.

The real part of the optical potential has been calculated by Sinha [24,25] using the double folding model and the Kallio-Kolltveit interaction, and taking account of the density dependence of the nucleon-nucleon interaction by including the linear factor:[26]

$$F(\rho) = \alpha(1 - \beta\rho) \quad , \tag{2.2}$$

where α,β are parameters and ρ is the total density at a point midway between the two interacting nucleons. This factor takes account of the saturation of the nuclear forces and thus reduces the potential in the nuclear interior, as shown in Fig. 2.4. The values of the volume integral $J/A_p A_T$, are remarkably constant from ^{16}O to ^{208}Pb and have mean

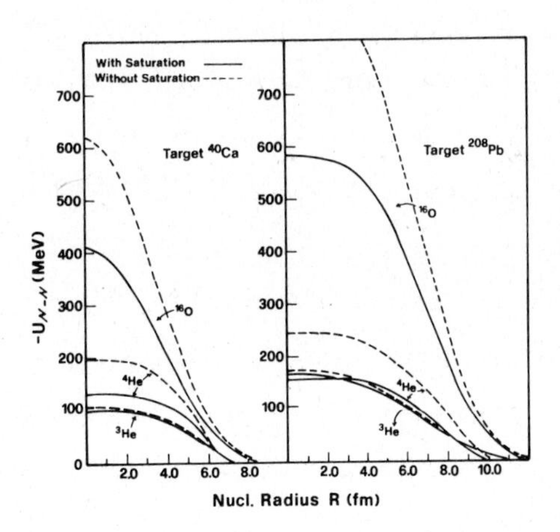

Fig. 2.4. Optical potentials calculated by a folding model with a density-dependent interaction showing the effect of saturation.[25]

values 333 and 371 for alpha-particles and helions respectively. The radius parameter $r_o=R(A_1^{1/3} + A_2^{1/3})^{-1}$ remained close to unity for both projectiles.

Sinha also evaluated the exchange term that arises from the antisymmetrisation of the coordinates of the target and projectile nucleons and found it to be less than 2% of the direct term, which is negligible in view of the uncertainties in the assumptions underlying the calculation. Since most of the energy-dependence of the potential comes from this exchange term, the energy dependence is also found to be small.

Sinha also calculated the imaginary part of the potential by first using the forward-scattering amplitude approximation[27)] to calculate the nucleon-nucleus imaginary potential taking account of both the internal motion and the relative external motion of the projectile nucleons and then folding this with the projectile density. Some of his results are given in Fig. 2.5.

These folding model potentials were compared with the differential cross-section for the elastic scattering of 141.7 MeV alpha-particles by ^{40}Ca and ^{90}Zr, and of 51.4 and 83.5 MeV helions by ^{40}Ca. The fits were similar to those obtained by phenomenological optical model analyses. In the case of the helion analysis it was necessary to adjust the depth or form of the imaginary potential to optimise the fit.

The folding model has been used by Perkin, Kobos and Rook[17)] to analyse the elastic scattering of alpha-particles by ^{90}Zr at several energies from 40 to 100 MeV. They found that it is not possible to obtain an acceptable fit with the folded potential alone, but if it is joined to a Saxon-Woods potential in the far surface region the resulting potential fits better than the best Saxon-Woods potential on its own, as

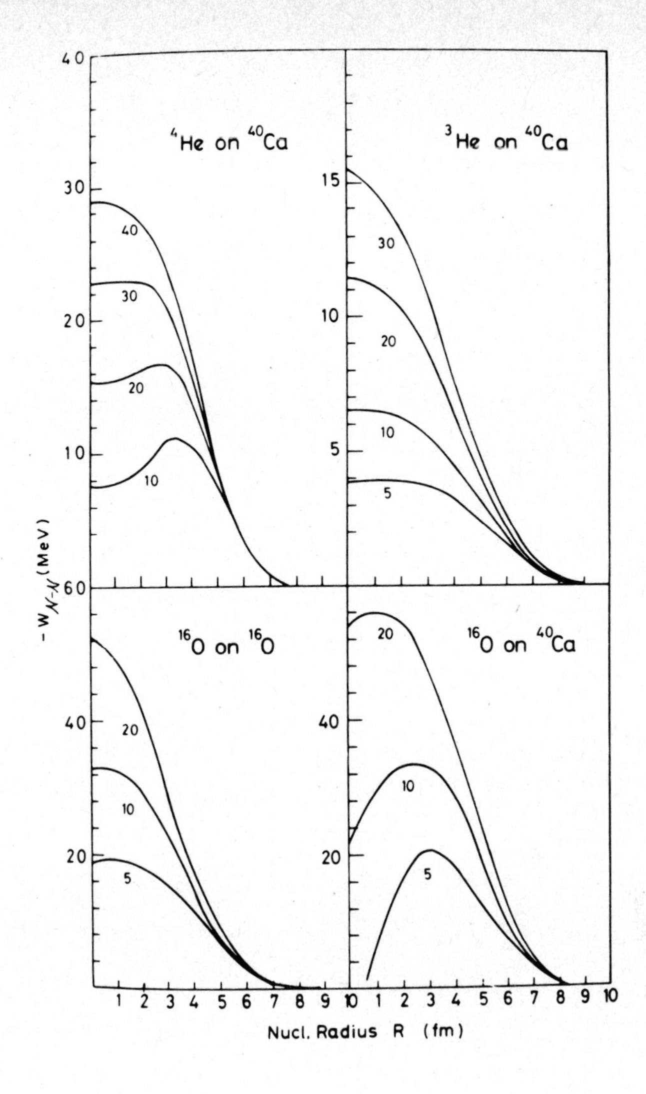

Fig. 2.5. Imaginary optical potentials calculated by a folding mo del for various particles. The numbers on the curve are the centre-of-mass energies per nucleon of the projecti les.[25)]

shown in Fig. 2.6. The matching radius was chosen as large as possible, providing the fit was still acceptable, and varied from 5.5 to 7.2 fm, compared with the charge radius of less than 5 fm. It is surprising that the folding model seems to fail in the far surface region; Perkin et al. suggest that this might indicate that the target nucleus is very much changed in this region by the presence of the projectile.

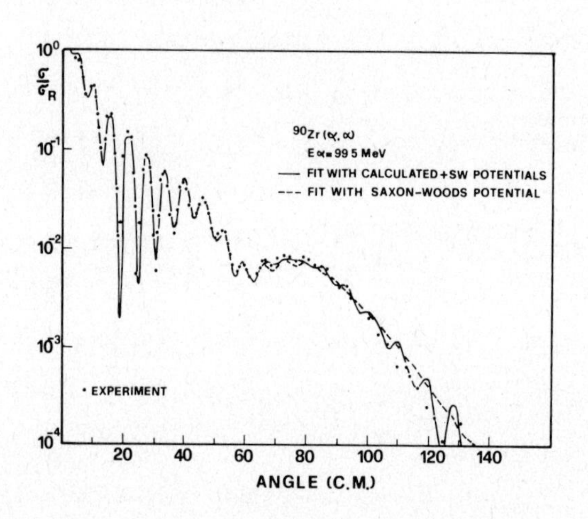

Fig. 2.6. Differential cross-section for the elastic scattering of 99.5 MeV alpha-particles by ^{90}Zr compared with optical model calculations using (a) calculated (folding model) + Saxon-Woods potentials (full curve) and (b) Saxon-Woods potential alone (dashed curve).[17)]

3. Density Distributions from Single-Particle States

The nuclear density distributions may be calculated from the simple shell model, in which each nucleon moves in a one-body potential. Once the potential is fixed, the wave functions of all the nucleons can be calculated, and the sum of their squared moduli gives the nucleon density distribution, and folding in the nucleon charge and matter distributions gives the nuclear charge and matter distributions.

Calculations by Donnelly and Walker[28] showed that charge distributions obtained in this way using an harmonic oscillator potential account quite well for the electron elastic scattering form factor in the region of the first minimum, but are unable to give the second minimum. Calculations with a Saxon-Woods potential were able to do this, in qualitative agreement with the experimental values.

The parameters of the potential may be fixed phenomenologically. If a Saxon-Woods form is used, the radius and diffuseness parameters may be chosen from a wide range of analyses that all give $r_o \simeq 1.25$ fm and $a \simeq 0.6$ fm. The depth of the potential is then adjusted to give the binding energy of each nucleon in turn, and suitable values are the centroid energies obtained from studies of one-nucleon transfer reactions. If a spin-orbit term is included in the potential, its value may be found from the separation in energy of the $J=L\pm1/2$ doublets.

Several density distributions have been calculated in this way,[28-30,9] and some of the results are shown in Fig. 3.1. In this case there is some uncertainty due to the nuclear deformation, but on the whole the agreement is sufficient to encourage further study.

One of the difficulties of these calculations is the determination of the binding energy of the single particle state. Most states are split into a number of fragments by the residual interactions, and the appropriate binding energy is their centroid, each fragment being weighted by its spectroscopic strength.[31] Estensive spectroscopic studies are thus necessary to determine the required binding energies for each nucleus.

A considerable simplification is however introduced by the systema-

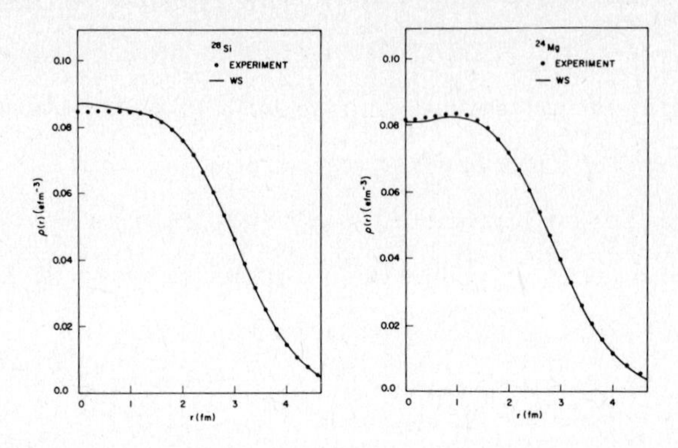

Fig. 3.1. Charge distribution for ^{24}Mg and ^{28}Si obtained from electron elastic scattering compared with SP calculations.[9)]

tic behaviour of the energies of these single-particle states from nucleus to nucleus;[32,33)] It is found possible to express the potential depth, for a fixed form factor, as a simple function of A, and the nuclear asymmetry parameter (N-Z)/A, for each state. This makes it possible to calculate the required binding energies to a high degree of accuracy even for nuclei for which no spectroscopic data are available.

This method is able to give the binding energies of states near the Fermi surface, but except for the very light nuclei it is not possible to determine those of the deeply-lying states. Fortunately the state dependence of the single-particle potential does not seem to be greater than about 10%, so it is sufficient to use for the deep states the average of the potentials found for those states accessible to spectroscopic investigation. The sensitivity studies described in the next section show that the uncertainty introduced in this way is very small.

Density distributions calculated in this way have been used to analyze nucleon and heavy ion scattering. Thus Kujawski and Vary[34] analyzed the elastic scattering of 1 GeV protons by ^{58}Ni and ^{208}Pb using the first order optical potential with a spin-orbit term, and some of their results are shown in Fig. 3.2.

A more detailed folding model of the heavy ion optical potential has been developed by Dover and Vary[35] using the expression

$$V(r) = \int \rho_A(\underline{r}_1)\rho_B(\underline{r}_2)G(\underline{r}+\underline{r}_1-\underline{r}_2)d\underline{r}_1\, d\underline{r}_2 \quad , \tag{3.1.}$$

where $G(\underline{r})$ is the effective nucleon-nucleon interaction that includes rescattering corrections to all orders with the restrictions due to the Pauli principle and other many-body effects. For computational convenience $G(\underline{r})$ was parametrised by the Gaussian form

$$G(r) = \bar{f}\, N\exp(-r^2/r_o^2) \quad , \tag{3.2.}$$

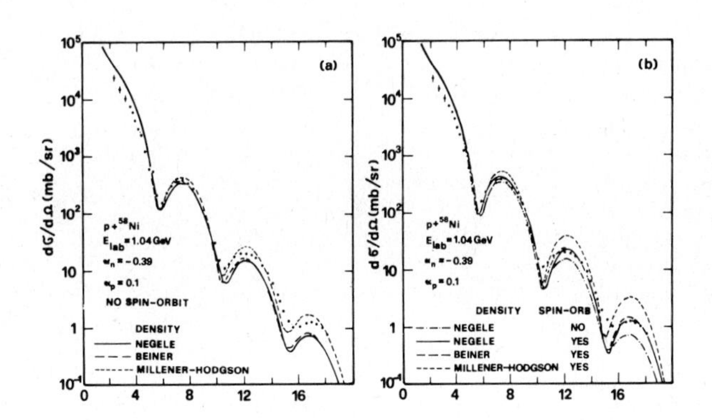

Fig. 3.2. Differential cross-section for the elastic scattering of 1 GeV protons by ^{58}Ni compared with optical model calculations obtained from SP densities.[34]

where $\overline{f}$ is a complex depth parameter and N a normalisation constant chosen so that $N\int \exp(-r^2/r_o^2)dr = -2\pi\hbar^2 M$. The nuclear densities were calculated from the single particle potentials of Millener and Hodgson.[32)]

The usefulness of this model can be tested by fitting experimental data and then comparing the optimum values of the parameters with the results of calculations based on our knowledge of the nucleon-nucleon interaction. Two fits to experimental data are shown in Fig. 3.3; their

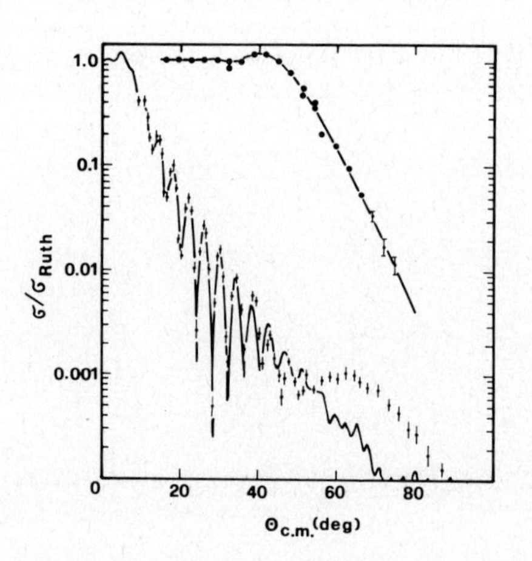

Fig. 3.3. Differential cross-sections for the elastic scattering of 62.92 MeV ^{18}O by ^{60}Ni and 139 MeV alpha-particles by ^{208}Pb compared with optical model calculations with doubly-folded potentials with strengths $\overline{f}$=1.27+0.9i and 1.79+1.21i respectively, and an effective nucleon-nucleon interaction range parameter of 1.4. fm.[35)]

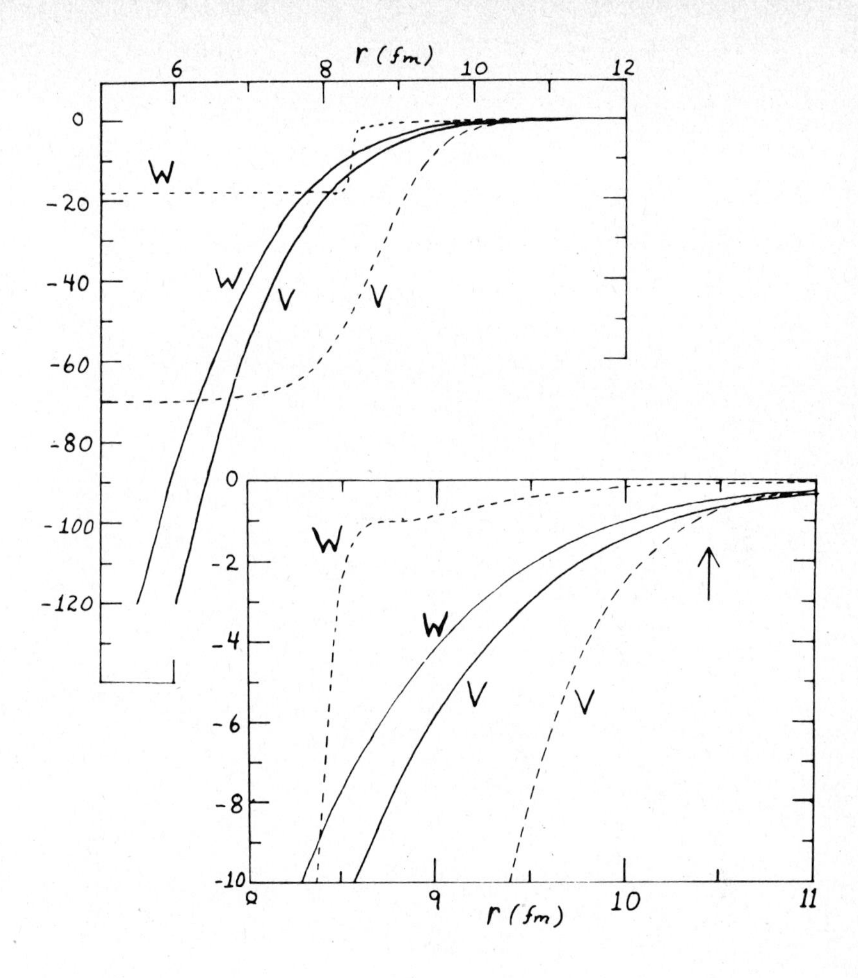

Fig.3.4. Real and imaginary parts of the folded potential correspon ding to the $^{18}O+^{60}Ni$ fit of Fig.3.3 (solid curves) compared with the best fit Saxon-Woods potential (dashed curves). The arrow indicates the region of greatest sensitivity for elastic scattering.[35)]

quality is comparable with those obtained with a Saxon-Woods potential with four or six parameters. The model thus provides a good description of the interaction at least in the surface region.

The potential corresponding to the fit to the $^{18}O + ^{60}Ni$ data are shown in Fig. 3.4, compared with the corresponding best fit Saxon-Woods potential.They differ very markedly from each other, except for the real part in the surface region indicated by the arrow. This shows once again the sensitivity of the elastic scattering to one small region of the potential, and also its general insensitivity to the imaginary part.

It now remains to connect the empirical values of $\bar{f}$ and r_o to the known characteristics of the nucleon-nucleon interaction. Neglecting many-body effects the simplest theoretical estimate for $\bar{f}$ is

$$\bar{f} = \sum_L \bar{f}_L \quad , \tag{3.3}$$

where $\bar{f}_L$ is the partial wave amplitude averaged over spin and isospin. Thus for S-waves

$$\bar{f}_o = (1- \tfrac{1}{2}\xi)\bar{f}(^1S_o) + \tfrac{3}{2}\xi\, \bar{f}(^3S_1) \quad , \tag{3.4}$$

where

$$\bar{f}(SLJ) = \frac{2L+1}{k}\exp(i\delta_{SLJ})\sin\delta_{SLJ} \quad ,$$

where δ_{SLJ} is the free space NN phase shift for spin S, orbital angular momentum L and total spin J. The factor ξ gives the proper spin-isospin average.

Dover and Vary calculated f_o from the on-shell phase-shifts for the s, p and d waves obtained from the two-body scattering data by MacGregor, Arndt and Wright,[36] and averaged over the spin and isospin stati

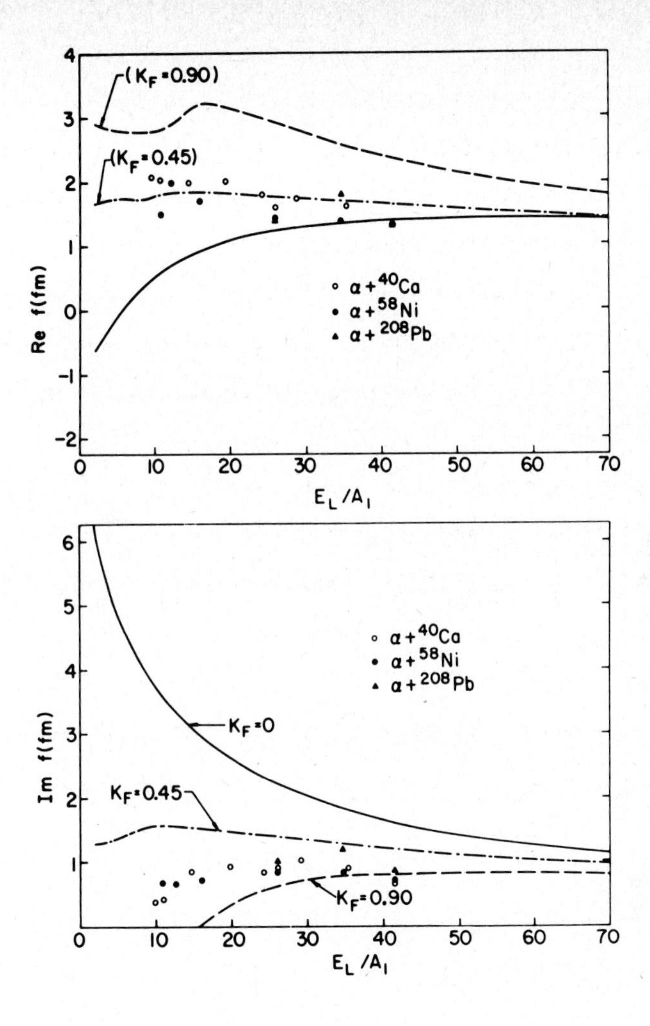

Fig.3.5. Energy dependence of the real and imaginary parts of the strength parameter $\overline{f}$ compared with values calculated from the nucleon-nucleon amplitudes for various values of the Fer mi momentum k_F, including the Pauli principle and off-shell effects.[35]

stics and over the Fermi motion. They also took the off-shell effects into account, and corrected for the Pauli principle by the reference spectrum method. The resulting values of $\overline{f}$ are compared with the best phenomenological values in Fig. 3.5 for several values of the Fermi momentum k_F. The optimum value is around 0.45 fm^{-1}. They consider that k_F should be in the range $0.45 \leq k_F \leq 0.90$ fm^{-1} corresponding to a local density in which the two nucleons collide in the range of 4% to 33% of the nuclear saturation density. Thus around 40 MeV per nucleon the model with $k_F = 0.45$ fm^{-1} gives the strength parameters to about 20%. The range parameter r_o was found phenomenologically to be around 1.40 ± 0.25 fm, which is consistent with what is known of the range of the effective nucleon-nucleon force.

In this work Dover and Vary initially used electron scattering densities, but found that this gave irregularities in the analysis of the ^{16}O and ^{18}O data. This is due to the strong sensitivity to the density in the tail region, which is not well determined by electron scattering. In later work they used the average single-particle potentials of Millener and Hodgson [32)] to generate the wave functions and hence the density distributions. These have a sounder physical basis, and are more reliable in the far surface region, which is the most important region for heavy ion interactions. Indeed, a calculation of the strong absorption radius shows that the main contribution to the optical potential comes from densities less than one-tenth of the central value. This shows that a low-density expansion for the heavy ion potential should be valid at low energies for peripheral interactions. The situation is quite different for the nucleon-nucleon potential, where the interaction is spread through the interior of the nucleus.

Since the elastic scattering is sensitive only to a very restricted

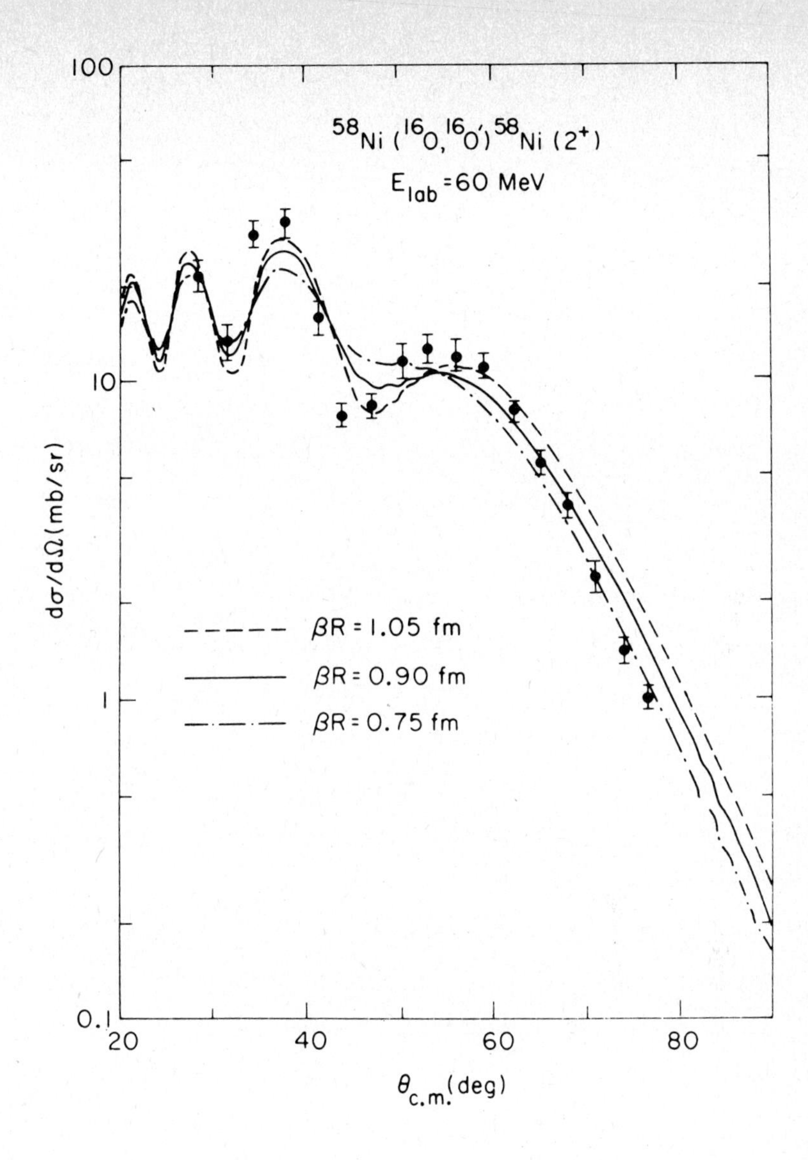

Fig. 3.6. Dependence of the calculated inelastic $^{58}Ni(^{16}O,\ ^{16}O')$ $^{58}Ni^{*}(2^+)$ cross section on the parameter $\beta_2 R_T$. The three calculated curve all use the same form factor but different values of $\beta_2 R_T$ as labelled.

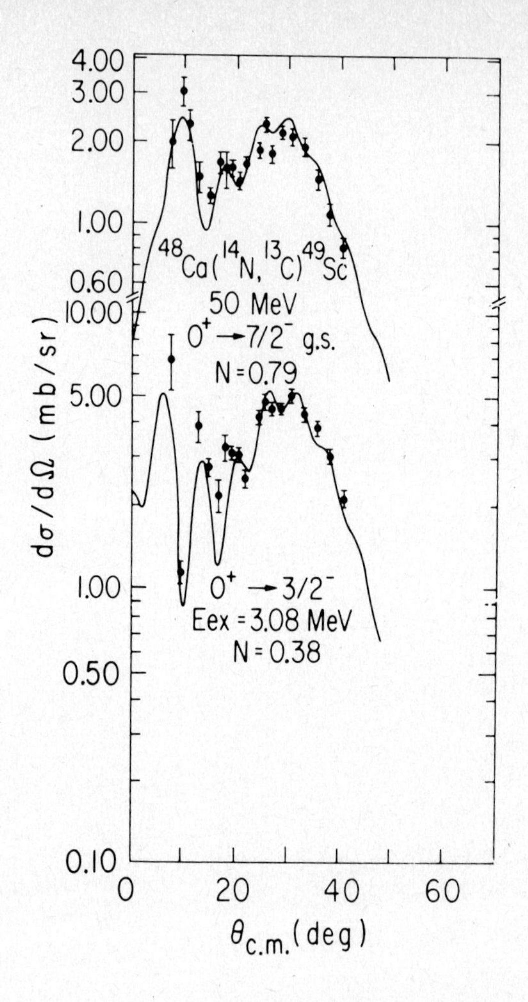

Fig. 3.7. One particle transfer cross sections obtained using the folded potential which best fits elastic scattering.

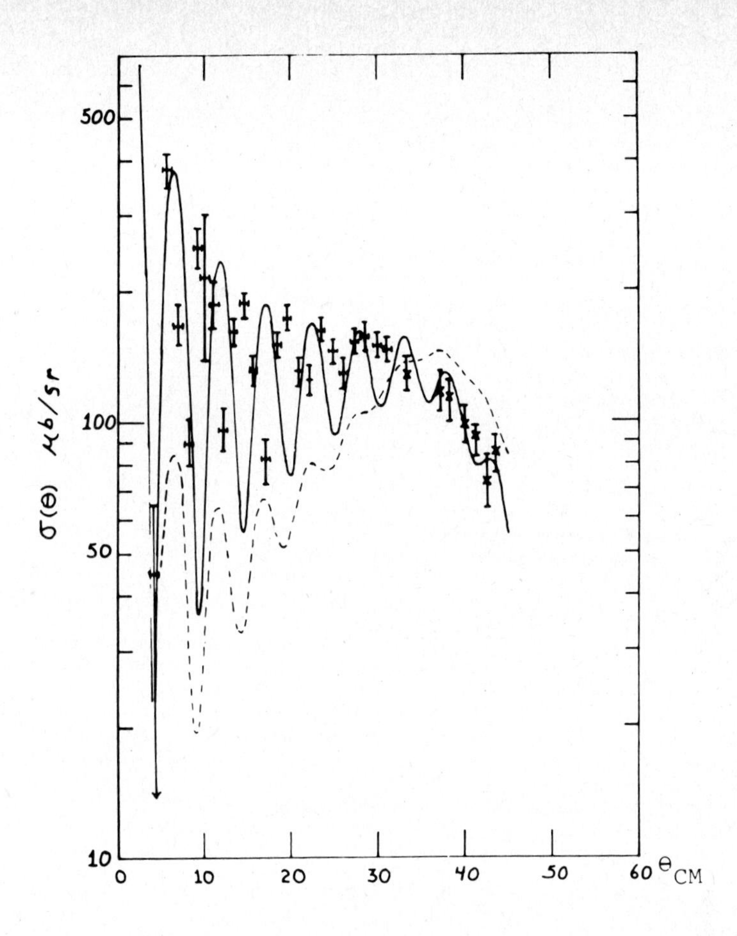

Fig.3.8. Two particle transfer cross section (dashed line) for the $^{60}Ni(^{18}O,^{16}O)^{62}Ni$ reaction at 62.92 MeV obtained using the best fit folded potential of Fig.3.4. Numerical densities obtained from a single particle model were used. The solid line shows the effect of decreasing the absorption by 50%; a reasonable fit to the elastic data is maintained.

region of the potential, it does not provide a very stringent test of the model. It has therefore been applied to inelastic scattering (Fig. 3.6), one-nucleon transfer reactions (Fig. 3.7) and multinucleon transfer reactions (Fig. 3.8), and on the whole has been found to give a good account of the experimental data. It has also been successfully applied to calculate the energies and widths of cluster states in light nuclei.[37)]

It was at one time hoped that the elastic scattering, inelastic scattering and nucleon transfer reactions would successively probe more deeply the potential. To test this, Moffa et al.[38)] made a series of calculations of the sensitivity of the goodness-of-fit to cutting off the potential above and below the sensitive region. They found for each type of interaction that the cross-section depends on the potential in a very limited radial range, and is quite insensitive to it outside this range. The ranges are shown in Fig. 3.9 and it is clear that all three interactions probe the potential to essentially the same minimum radius, wich is just inside the strong absorption radius.

It might, in addition, be hoped that measurement of the cross-sections to very low intensities and high angles might probe the potential more deeply. Since low intensities can most easily be measured for elastic scattering, the effect of including cross-sections down to 10^{-4} of the Rutherford value was also studied by Moffa et al.[38)] Their results, shown in Fig. 3.10, confirm the expectation that the cross-sections at higher angles give information about deeper regions of the potential, but the extra depth probed by a further decade of intensity of cross-section is small and decreasing.

In the light of these results, it is not surprising that the inelastic and transfer data are also well fitted by a heavy ion potential

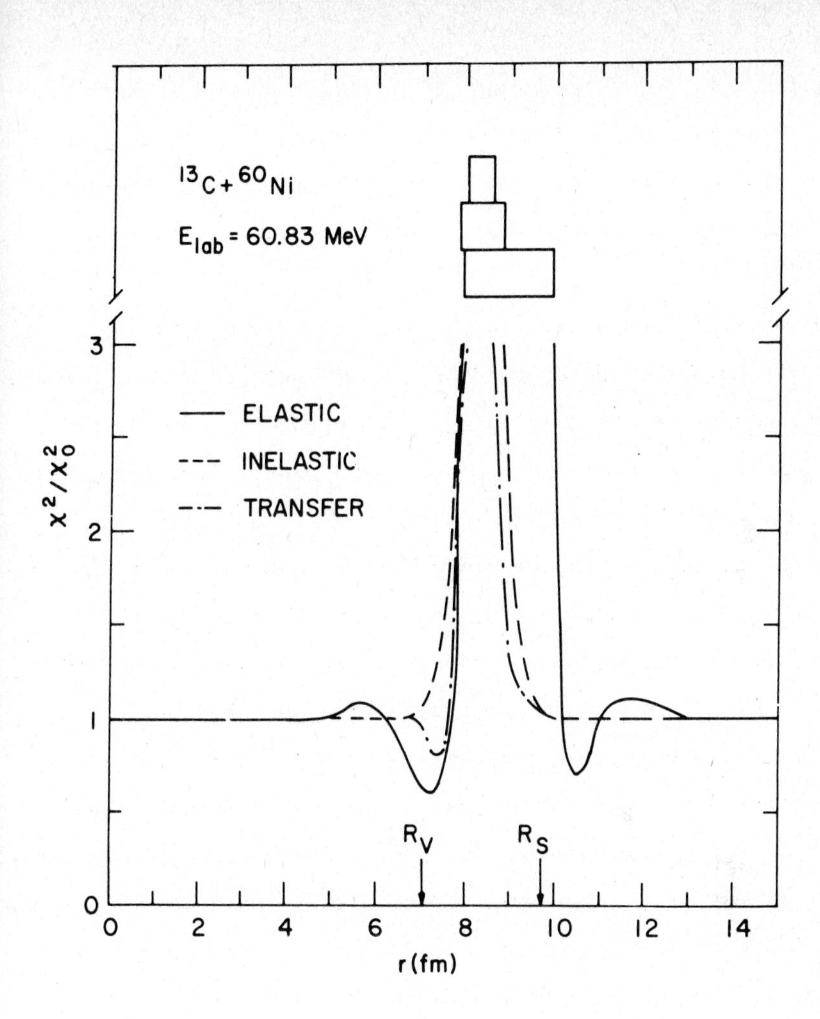

Fig.3.9. Regions of sensitivity of the elastic scattering, inelastic scattering and one-neutron transfer reactions to the real part of the optical potential. The radius of the potential is R_V and the strong absorption radius is R_S.[38)]

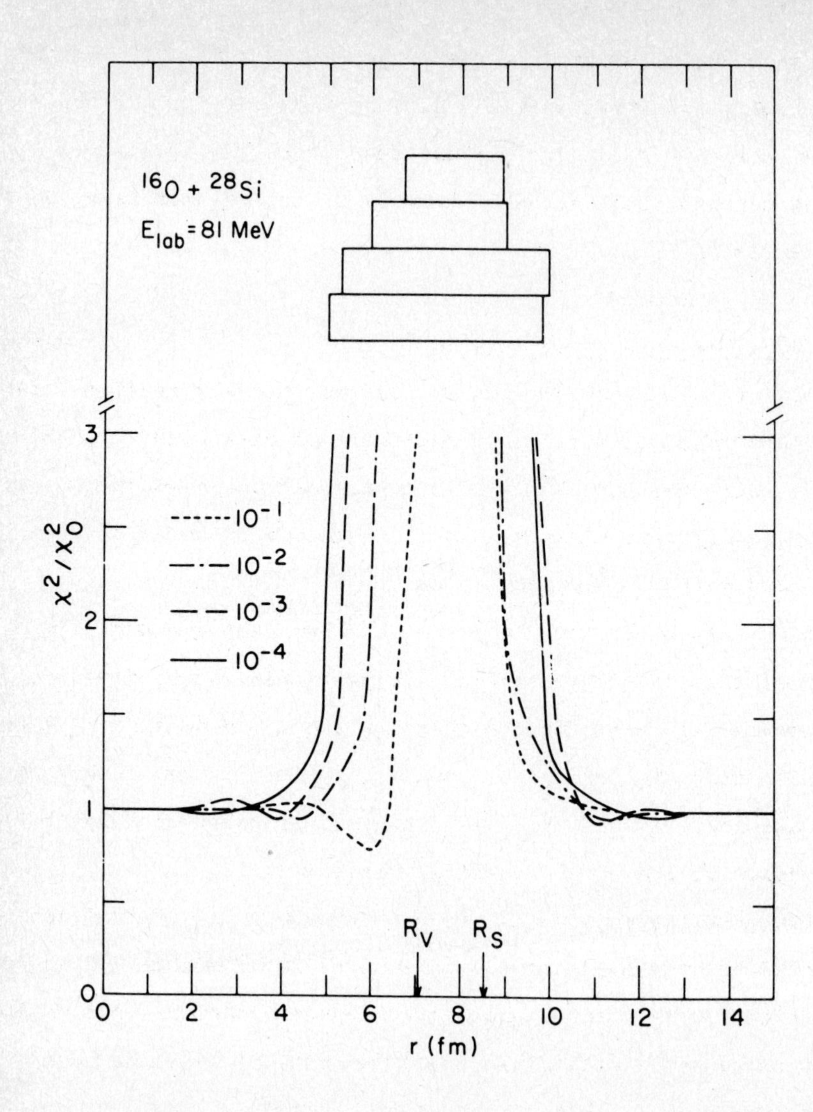

Fig.3.10 Regions of sensitivity of the elastic scattering cross-section to the real part of the optical potential as a function of the range of intensities (expressed as a ratio to Rutherford cross-sections) for which the cross-section is available.[38)]

hat fits the elastic scattering. The regions of sensitivity are all utside the radius of the potential, showing that all these interacions are essentially peripheral. These results also suggest that not uch meaning can as yet be attached to volume integrals of heavy ion ptical potentials.

. Improved Nuclear Density Calculations

The success of the SP method for calculating nuclear densities,as hown by the improved fits to reactions that are very sensitive to the ensity (and hence the potential) in the surface region encourages further effort to improve them by taking into account higher order efects that have hitherto been neglected.

In the first place, the single particle potentials are known only or the states near the Fermi surface. In previous calculations, an verage potential was used for the deeper states, and subsidiary calculations showed that the uncertainty introduced in this way is not likely to be critical. It can only be removed by an understanding of the state dependence of the single-particle potential.

The single-particle potentials can be taken from the general formulae, or fitted to the measured centroid energy in each case. It is probably better to use the latter procedure whenever possible, and to use the general formulae only when the requisite data is lacking. Again, sensitivity studies show that this is not critical.

The simple model assumes that the shell model orbits are fully occupied up to the Fermi level, and are thereafter empty, whereas it is well known that the occupation probabilities are somewhat less than unity for several levels near the Fermi surface, and are non-zero for states normally considered to be unoccupied. This may easily be included in the calculation simply by weighting the charge distributions corresponding to particles in each orbit by the corresponding occupation probabilities. This was first done by Elton, Webb and Barrett[39] and by Elton and Webb.[40]

In the work of Li et al.[9] the 2s1d occupation number was treated as an adjustable parameter, and without this flexibility it was not possible to obtain an acceptable fit. They found 2s occupation numbers of 0.6, 0.9 and 1.4 for ^{24}Mg, ^{28}Si and ^{32}S, compared with values of (0.19,0.46), 0.79 and 1.5 found from analyses of stripping and pickup reactions. This shows the consistency of the calculations and the importance of taking account of occupation probabilities. These calculations have been repeated and the results are shown in Figs.4.1-4.3.

One of the arbitrary features of the SP method is the choice of the form factor parameters $\underline{r}_o$ and $\underline{a}$ of the one-body Saxon-Woods potential. These are certainly known quite well, and the values chosen in the work of Millener and Hodgson[32] are confirmed by the agreement, within statistical uncertainties, between the RMS radii of the SP charge distributions and those obtained from analysis of electron scattering and muonic atom data (EE distributions). Some improvement may be effected, however, by requiring that they give the same RMS radius as the EE distribution. It might also be possible to use the $\langle r^4 \rangle$ of the EE distributions to provide a further constraint on the form factor parameters.

This adjustment of the form factor parameters can be made only for the charge distribution, i.e. for the proton potentials. It was assumed that the neutron potential parameters are the same. This appears to be in accord with the much less accurate data on nuclear matter distributions.

The value of the folded second and fourth moments of the charge,

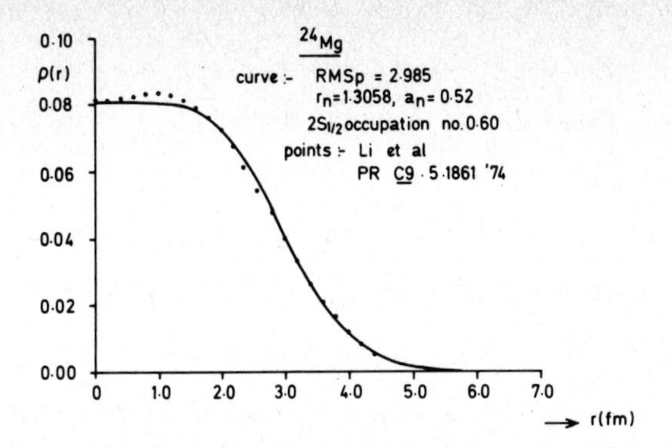

Fig. 4.1. Calculated WS density of ^{24}Mg compared with the experimental results of Li et al.[9)]

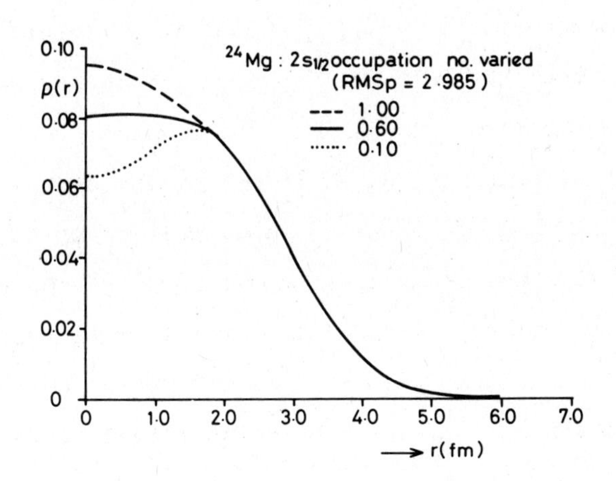

Fig. 4.2. Calculated density of ^{24}Mg showing the effect of varying the $2s_{1/2}$ occupation number.

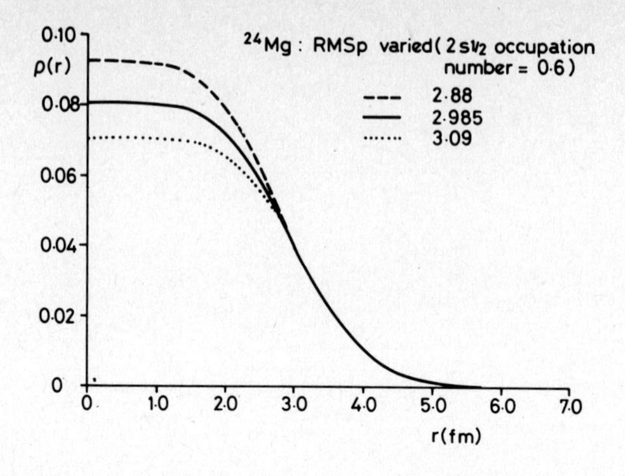

Fig. 4.3. Calculated density of ^{24}Mg showing the effect of varying the RMS charge radius.

distribution can be calculated directly from the unfolded values using the relations

$$\langle r^2\rangle_{ch} = \langle r_p^2\rangle + \langle r^2\rangle \quad , \tag{4.1}$$

and

$$\langle r^4\rangle_{ch} = \langle r_p^4\rangle + \langle r^4\rangle + (10/3)\langle r_p^2\rangle\langle r^2\rangle \quad , \tag{4.2}$$

where $\langle r_p^2\rangle$ and $\langle r_p{}^4\rangle$ refer to the nuclear proton density distribution and $\langle r^2\rangle$ and $\langle r^4\rangle$ to the charge distribution of the proton itself. For $\langle r^2\rangle$ we use the value 0.775 corresponding to a best fit with a sum of three Gaussians; the corresponding value of $\langle r^4\rangle$ is 1.824 (Ref.41).

There are several small corrections to these moments that should be considered, including the effect of the charge distributions of the neutrons and relativistic and spin-orbit effects.

Although their total charge is zero, neutrons have a charge distribution with finite moments, and this is sufficiently important to be included in any calculation of the nuclear charge distribution. It was indeed found by Bertozzi et al.[42] to be sufficient to show that the apparently anomalous decrease of the RMS charge radius from ^{40}Ca to ^{48}Ca is due to the $f_{7/2}$ neutrons. The mean square neutron charge radius is about $-0.116\ fm^2$ (Ref. 43), so the total mean square radius can be obtained by subtracting 0.116 N/Z from 0.775. The effect on the charge distribution itself may be obtained by adding

$$V_N(r) = \int \rho_N(\underline{r}')\rho_n(|\underline{r}-\underline{r}'|)d\underline{r}' \tag{4.3}$$

to that already calculated from the proton distribution.

Relativistic corrections to the charge distribution arise due to the non-relativistic reduction of the four-component single nucleon current to a two-component form.[42,44] For heavy nuclei this correction is almost exactly cancelled by the correction for the spurious centre-of-mass motion.[41]

The correction due to the spin-orbit forces was calculated by Chandra and Sauer and found to be substantially less than that due to the neutron charge distribution (Fig. 4.4). In ^{208}Pb, the spin-orbit contributions from spin unsaturated neutrons and protons almost exactly cancel.[42]

Several calculations have been made to explore the usefulness of the SP method. The charge distribution of ^{208}Pb has recently been determined by Euteneuer et al.[45] from electron elastic scattering. Their distribution is compared with the SP distribution in Fig. 4.5. The radius parameter r_o was fixed to give the experimental RMS charge radius of 5.491 fm, and the corresponding value of $\langle r^4 \rangle$ is 34.021 compared

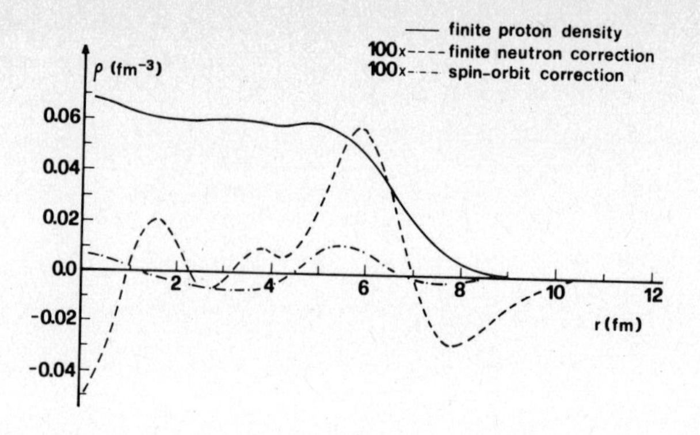

Fig. 4.4. Corrections to the calculated charge density of ^{208}Pb due to the finite proton density (full curve), finite neutron size (dashed curve) and spin-orbit forces (dot-dash curve). The last two curves have been multiplied by one hundred.[41)]

with the experimental value of 34.034. It was not possible to make these values coincide by choice of diffuseness parameter a, as was hoped because the loci of the experimental RMS and RMF radii do not intersect. Subsidiary calculations showed that their separation is sensitive to quite small corrections like the value chosen for the neutron RMS radius, and that the resulting charge distribution is very insensitive to r_o and a.

Using the same parameters, the charge density difference between ^{208}Pb and ^{209}Bi was calculated, and is compared in Fig. 4.6 with the experimental results of Sick.[46)]

It is also necessary to see the effect of using non-local wave functions in place of the equivalent local ones, and studies of this are in progress.

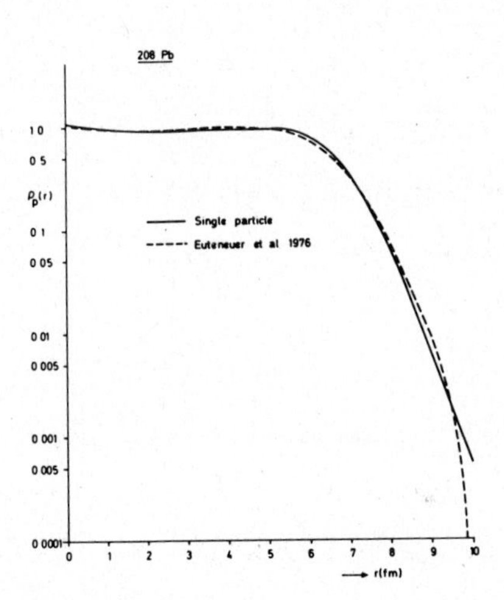

Fig. 4.5. Calculated charge distribution of ^{208}Pb compared with SP distribution.[45)]

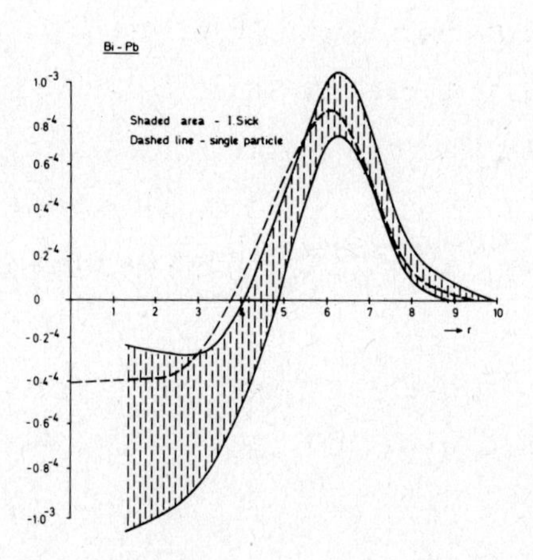

Fig. 4.6. Calculated charge density difference between ^{208}Pb and ^{209}Bi compared with the experimental results of Sick.[7)]

References

1) G.W.Greenlees, G.H.Pyle and Y.C.Tang, Phys. Rev. 171, 1115 (1968).

2) P.E.Hodgson, Nuclear Reactions and Nuclear Structure (Oxford 1971) Ch. 2.

3) J.Friedrich and F.Lenz, Nucl. Phys. A183, 523 (1972).

4) J.L. Friar and J.W.Negele, Nucl. Phys. A212, 93 (1973).

5) J.Borysowicz and J.H.Hetherington, Phys. Rev. C7, 2293 (1973).

6) J.H.Hetherington and J.Borysowicz, Nucl.Phys. A219, 221 (1974).

7) I.Sick, Nucl. Phys. A218, 509 (1974).

8) I.Sick, J.B.Bellicard, M.Bernheim, B.Frois, M.Huet, Ph.Leconte, J.Mougey, Phan Xuan-Hô, D.Royer and S.Turck, Phys. Rev. Lett.35, 910 (1975).

9) C.G.Li, M.R.Yearian, and I.Sick, Phys. Rev. C9, 1861 (1974).

10) J.P.Vary and C.B.Dover, Phys. Rev. Lett. 31, 1511 (1973).

11) C.B.Dover and J.P.Vary, Symposium on Classical and Quantum Mechanical Aspects of Heavy Ion Collisions (Springer-Verlag 1975).

12) C.J.Batty, E.Friedman and D.F.Jackson, Nucl. Phys. A175, 1 (1971).

13) A.Budzanowski, A.Dudek, K.Grotowski, Z.Majka and A.Strzalkowski, Part. & Nucl. 6, 97 (1973).

14) J.S.McIntosh, S.C.Park and G.H.Rawitscher, Phys. Rev. 134, B1010 (1964).

15) G.R.Satchler, Nashville Conference, II. 171 (1974).

16) Y.Eisen, Phys. Lett. 37B, 33 (1971).

17) D.G.Perkin, A.M.Kobos and J.R.Rook, Nucl. Phys. A245, 343 (1975).

18) F.G.Perey and G.R.Satchler, Nucl. Phys. A97, 515 (1967).

19) B.Block and F.B.Malik, Phys. Rev. Lett. 19, 239 (1967)

20) R.H.Siemssen, J.V.Maher, A.Weidinger and D.A.Bromley, Phys. Rev.

Lett. 19, 968 (1967).

21) P.G.Zint and U.Mosel, Phys. Lett. 56B, 424 (1975).

22) S.M.Perez, Phys. Rev. C8, 1606 (1973).

23) N.Rowley, Thesis (Oxford, 1973).

24) B.Sinha, Phys. Rev.Lett. 33, 600, (1974).

25) B.Sinha, Phys. Rev. C11, 1546 (1975).

26) A.Lande, A.Molinari and G.E.Brown, Nuclear Phys. A115, 241 (1968).

27) B.Sinha and F.Duggan, Nucl. Phys. A226, 31 (1974).

28) T.W.Donnelly and C.E.Walker, Phys. Rev. Lett. 22, 1121 (1969).

29) L.R.B.Elton and A.Swift, Nucl. Phys. A94, 52 (1967).

30) B.B.P. Sinha, G.A.Peterson, R.R.Whitney, I.Sick and J.S.McCarthy, Phys. Rev. C7, 1930 (1973).

31) P.E.Hodgson, Rep. Prog. Phys. 38, 847 (1975).

32) D.J.Millener and P.E.Hodgson, Nucl.Phys. A209, 59 (1973).

33) F.Malaguti and P.E.Hodgson, Nucl. Phys. A215, 243 (1973);A257, 37 (1976).

34) E.Kujawski and J.P.Vary, Phys. Rev. C12, 1271 (1975).

35) J.P.Vary and C.B.Dover, Second High Energy Heavy Ion Summer Study, Lawrence Berkeley Laboratory (1974); C.B.Dover, P.J.Moffa and J.P.Vary, Phys. Lett. 56B, 4 (1975).

36) M.H.MacGregor, R.A.Arndt and R.M.Wright, Phys. Rev. 182, 1714 (1969).

37) B.Buck, C.B.Dover and J.P. Vary, Phys.Rev. C11, 1803 (1975).

38) P.J.Moffa, C.B.Dover and J.P.Vary, Phys. Rev. C13, 147 (1976).

39) L.R.B.Elton, S.J.Webb and R.C. Barrett, Proc. Third International Conf. on Nuclear Structure, New York (1969).

40) L.R.B. Elton and S.J.Webb, Phys. Rev. Lett. 24, 145 (1970).

41) H.Chandra and G.Sauer, Phys. Rev. C13, 245 (1976).

42) W.Bertozzi, L.Friar, L.Heisenberg and J.W.Negele, Phys. Lett. 41B, 408 (1972).

43) F.A.Bumiller, F.R.Buskirk, J.W.Stewart and E.B.Dally, Phys. Rev. Lett. 25, 1774 (1970).

44) T.de Forest and J.D.Walecka, Adv. Phys. 15, 1 (1966).

45) H.Euteneuer, J.Friedrich and N.Voegler, Phys. Rev. Lett. 36, 129 (1976).

46) I.Sick, Nucl. Phys. A208, 357 (1973).

NON-LOCAL OPTICAL POTENTIAL: THEORETICAL AND PHENOMENOLOGICAL ASPECTS†

M.M. GIANNINI AND G. RICCO

Istituto di Scienze Fisiche dell'Università-Genova

Istituto Nazionale di Fisica Nucleare-Sezione di Genova

Abstract: The general expression of the nucleon-nucleus optical potential has been obtained using Watson's multiple scattering theory and Wolfenstein's parametrization of the nucleon-nucleon scattering amplitude. The resulting theoretical potential is non-local and consists of an energy independent central volume plus surface real and imaginary potential and of a Thomas-like spin-orbit term. The analysis has been restricted to N=Z spherical nuclei, so that neither isospin-isospin nor spin-spin interaction have been included.
The widely used Perey-Buck, Greenlees and Watson expressions of the optical potential are easily obtained as particular cases. For practical purposes the non-local potential has been parametrized in the Frahn-Lemmer form, using Woods-Saxon radial form factors and the equivalent local potential (ELP) has been calculated by a Perey-Buck-like transformation.

The ELP has a radial behaviour very similar to the original non-local one, but the potential depths and radii are energy dependent. The six free parameters in the ELP have been adjusted to fit the available experimental data in the -70÷ +150 MeV range of interest in nuclear reactions, namely energies of single hole and single particle states, charge distributions, proton elastic scattering cross sections and po-

† Presented by M.M. Giannini.

larizations. The fitted potential depths show an energy dependence in remarkable agreement with the model predictions with a central non-locality range $\beta \simeq 1.$fm and a spin-orbit non-locality range $\beta_s \equiv 0.8$ fm. The relative importance of surface and volume dependence in the real central potential in also discussed.

1. Introduction

There are mainly two different ways of using the optical potential in the analysis of the elastic nucleon-nucleus scattering data. The first one is purely phenomenological. The potential is assumed to have a real central and spin-orbit part and an imaginary central term, which are all represented by means of Woods-Saxon functions or their derivate: the unspecified parameters (potential depths, radii and diffuseness) are then determined by fitting the elastic differential cross sections and polarizations.[1,2]

The analysis is performed for each target nucleus and each incident nucleon energy quite independently; the parameters can in principle be completely different in each case and the physical meaning of the resulting potential is somewhat dubious. Moreover, there are some ambiguities in the parameters, such as the famous $V R^2$ = const, so the fitting procedure doesn't determine the potential univoquely.

The second approach is theoretical. In the framework of the general description of the nucleon-nucleus scattering process an abstract operator is defined, whose matrix elements in the elastic channel are identified with the optical potential.[3] According to the technique used, different expressions are obtained, which can all be used in order to calculate explicitly the optical potential, provided that either the nucleon-nucleon interaction or the nucleon-nucleon scattering matrix or the nuclear wave function is available. So this fundamental approach, even if successful,[4] is based on the detailed knowledge of quantities, which cannot be univoquely extracted from the present experimental data.

An intermediate point of view can be adopted. For instance, in the case of Greenlees approach,[5] a simple model is formulated for the local optical potential:

$$V(\vec{r}) = \int d\vec{r} \quad \rho_m(\vec{r}') \, \mathcal{V}(|\vec{r} - \vec{r}'|)$$

where $\rho_m(\vec{r}')$ is the nuclear matter density and $\mathcal{V}(x)$ is the nucleon-nucleon interaction.

Taking into account the most general expression of $\mathcal{V}(x)$, the optical potential is given, for zero-spin target nuclei, by a sum of a central plus spin-and isospin-dependent terms. The nuclear matter distribution is assumed to have a Woods-Saxon form and the various interactions are taken of a Yukawa type. Then, adding an imaginary part, eight parameters are left free, which are determined by a fitting procedure of the experimental scattering data.

This intermediate approach can be further developed.[6] The most general theoretical optical potential is deducible from Watson multiple scattering theory.[7] The theoretical expression, simplified by the introduction of some approximations, is then used not to calculate the optical potential, but rather to determine the terms which are relevant for the description of the nucleon-nucleus interaction and to establish some correlation among the parameters of the potential.

The phenomenological analysis can now be performed with a reduced number of free parameters. However, the above mentioned ambiguities are still present. In order to avoid them, it is convenient to fit both scattering and bound state data.[8] So the optical parameters should be determined consistently with nucleon-nucleus elastic scattering and with single particle binding energies and charge distributions. The analysis has been restricted, in this work, to spherical nuclei between ^{12}C and ^{40}Ca, up to the π-meson threshold.

2. Theoretical aspects

The generalized optical operator U is given, in Watson theory, by a multiple-scattering expansion

$$(1) \qquad U = \sum_{j=1}^{A-1} \tau_j + \sum_{j\neq k,1}^{A-1} \tau_j \;(E^{\pm}H_o)^{-1}(1-P)\tau_k - \sum_{j=1}^{A-1} \tau_j (E^{+}-H_o)^{-1} P\tau_j + \ldots,$$

where τ_j is the single-scattering t-matrix for bound nucleons, P is the projection operator on the elastic channel; H_o is the target nucleus hamiltonian plus the incident kinetic energy. Two approximations can be introduced: 1) the impulse approximation: τ_j is substituted with the free nucleon-nucleon scattering matrix t_j; 2) the single-scattering approximation: only the first term in Eq. (1) is retained. Thus:

$$(2) \qquad U \cong \sum_{j=1}^{A-1} t_j \quad .$$

Eq. (2) is valid if the energy E is sufficiently high. However, Eq. (2) is used only to derive the form of the optical potential, so that we can expect that the conclusion which can be drawn from it should be true at low energy too.

The projection operator P has the form:

$$P = \sum_{\nu_t} \int d\vec{p} \;|\; g_o\vec{p}\,\nu_t> \;\; <g_o\vec{p}\nu_t \;| \;\equiv$$

$$\equiv \sum_{\nu_t,\nu} \int |\vec{p}_o\;\nu> \;|g_o\vec{p}\nu_t> \;\; d\vec{p}\;d\vec{p}_o \;<g_o\vec{p}\;\nu_t| \;\; <\vec{p}_o\nu| \quad ,$$

where $|\vec{p}_o \nu\rangle$ is the incident plane wave with momentum $\vec{p}_o$ and spin third component ν; $|g_o \vec{p}\, \nu_t\rangle$ is the target nucleus state vector, which is specified by the total momentum $\vec{p}$ and the spin third component ν_t; g_o indicates the nuclear intrinsic ground state wave function and is translation invariant. The matrix elements of U in the momentum and spin representation are

$$(3) \qquad \sum_{j=1}^{A-1} \langle \vec{p}_o\, \nu;\, g_o \vec{p} \nu_t |\, t_j | \vec{p}_o{}'\, \nu';\, g_o \vec{p}\,'\, \nu_t' \rangle \quad ;$$

the optical potential in the coordinate representation is obtained by a Fourier transformation with respect to the four variables $\vec{p}_o$, $\vec{p}$, $\vec{p}_o{}'$, $\vec{p}\,'$. The integrations can be reduced to two if we consider that t_j acts only on the coordinates of the j-th nucleon and is Galilean invariant, that is:

$$(4) \qquad \langle \vec{p}_o \nu;\, \vec{p}_j \nu_j\, | t_j | \vec{p}_o{}' \nu'\, ;\, \vec{p}_j{}' \nu_j' \rangle = \delta(\vec{K}-\vec{K}')\; t_j^{(\nu)}(\vec{k},\vec{k}') \quad ,$$

where $\vec{K} = \vec{p}_o + \vec{p}$; $\vec{k} = \frac{1}{2}(\vec{p}_o - \vec{p}_j)$ (and analogously for the primed quantities); (ν) is a shorthand notation for the spin indexes. This property is important, since it avoids to introduce at this point the approximation of neglecting the momentum of the nucleons within the target. Moreover, Eq. (4), together with the translation invariance of the ground state wave function g_o, ensures that the optical potential is local in the overall centre of mass coordinate:

$$\vec{G} = \frac{1}{A}\,(\vec{r}_o + (A-1)\,\vec{R})$$

($\vec{r}_o$ is the position of the incident particle, $\vec{R}$ that of the nuclear center of mass) and independent of it:

$$\langle \vec{r},\vec{R} | \mathcal{V} |\, \vec{r}\,',\, \vec{R}' \rangle \;=\; \delta(\vec{G}-\vec{G}')\; \mathcal{V}(\vec{r},\vec{r}\,') \quad ,$$

where $\vec{r} = \vec{r}_o - \vec{R}$.

If the Fourier transform of Eq. (3) is performed, then:

$$\mathcal{V}(\vec{r},\vec{r}')=(2\pi)^{-3}\sum_{j=1}^{A-1}\sum_{\nu_j\nu_j'}\int d\vec{p}\,d\vec{q}\,\exp(i\,\vec{p}.(\vec{r}-\vec{r}'))\,\exp(i\vec{q}.(\vec{r}+\vec{r}')/2)\,. \quad (5)$$

$$.\;t_j^{(\nu)}\left((1-\varepsilon/2)\,\vec{p}+\vec{q}/2;\;(1-\varepsilon/2)\vec{p}-\vec{q}/2\right)F_j^{(\overline{\nu})}(\vec{q};\vec{r}-\vec{r}')$$

($\varepsilon = (A-2)/(A-1)$), where F is a generalized form factor:

$$F_j^{(\overline{\nu})}(\vec{q};\vec{r}-\vec{r}') = \int d\vec{x}\,\exp(-i\vec{q}.\vec{x})\,\rho_j^{(\overline{\nu})}(\vec{x};\vec{r}-\vec{r}')$$

with

$$\rho_j^{(\overline{\nu})}(\vec{x};\vec{r}-\vec{r}') = \int \prod_{i=1}^{A-1} d\vec{z}_i\,\delta(1/(A-1)\cdot\sum_{k=1}^{A-1}\vec{z}_k)\,\delta(\vec{x}-\vec{z}_j)\quad .$$

$$.\;g^{*}_{o,\nu_t\nu_j}(\vec{z}_1,\ldots,\vec{z}_j-(A-1)/2A(\vec{r}-\vec{r}'),\ldots\vec{z}_{A-1})\,g_{o,\nu_t'\nu_j'}(\vec{z}_1,\ldots,\vec{z}_j+$$

$$+(A-1)/2A(\vec{r}-\vec{r}')\ldots.,\vec{z}_{A-1})\;.$$

$g_{o,\nu_t\nu_j}$ is the ground state wave function projected on the state in which the j-nucleon has third spin component ν_j.

From Eq. (5) we see that the optical potential is non-local, as it should be;[9)] the origin of the non-locality is, in this approach, the fact that the scattering t-matrix is completely off-shell, that is it depends on $\vec{p}$: if t were a function of $\vec{q}$ only, then the integration over $\vec{p}$ would lead to a $\delta(\vec{r}-\vec{r}')$.There is a second cause of non-locality, namely the presence of $\vec{r}-\vec{r}'$ in the form factor; it can be shown that this $\vec{r}-\vec{r}'$ dependence is due to the internal motion of the target nucleons. Of course there would be a third source of non-locality: if the antisymmetrization of the incident particle with the target ones is taken into account, the optical potential must contain a non-local exchange term. The absence of an explicit term of this kind is a typical problem of the multiple-scattering approach and is presumably taken into account implicitly by a "spurious" off-shell behaviour of the scattering t-matrix.

Eq. (5) shows that the optical potential has, in general, an intrinsic energy dependence, which is due, in this formulation, to the energy dependence of the nucleon-nucleon scattering amplitude.

The potential given in Eq. (5) is quite general and reproduces, with some further assumptions, many phenomenological potentials which are widely used.

Let us restrict ourselves to the diagonal part of Eq. (5)

$$\mathcal{V}_c(\vec{r},\vec{r}\,') = (2\pi)^{-3}\ (A-1) \int d\vec{p}\ d\vec{q}\ \exp(i\vec{p}.(\vec{r}-\vec{r}\,'))\exp(i\ \vec{q}/2.(\vec{r}+\vec{r}\,')) \\ .\ t\ ((\vec{p}+\vec{q})/2;\ (\vec{p}-\vec{q})/2\ )\ F(\vec{q};\vec{r}-\vec{r}\,') \tag{6}$$

(ε is put equal to 1).

If $t=t_E(q)$ does not depend on $\vec{p}$, then the potential is local:

$$\mathcal{V}_c(\vec{r},\vec{r}\,') = \delta(\vec{r}-\vec{r}\,')\ \ \mathcal{V}_c(r) \quad , \quad r = |\vec{r}| \ ,$$

with

$$\mathcal{V}_c(r) = (A-1) \int d\vec{q}\ \exp\ (i\vec{q}.\vec{r})\ t_E(q)\ F(\vec{q})\ \ ,\ F(\vec{q})=F(\vec{q};o) \quad ;$$

the optical potential is so expressed in terms of an "effective interaction". Furthermore, if the Born approximation is valid, we get the Greenlees folding rule:

$$\mathcal{V}_c(r) = (A-1) \int d\vec{q}\ \exp(i\ \vec{q}.\vec{r})\ \mathcal{V}(q)\ F(q) = (A-1)\int d\vec{x}\rho(\vec{x})\ v(|\vec{r}-\vec{x}|) \ ;$$

v is the nucleon-nucleon interaction. If, on the other hand, $F(q)$ has a forward peak:

$$\mathcal{V}_c(r) = (A-1)\ t_E(o) \int d\vec{q}\ \exp\ (i\ \vec{q}.\vec{r})F(q) = (A-1)(2\pi)^3 t_E(o)\rho(r) \quad ,$$

which is Watson's formula.

3. A non-local potential model

The preceding expressions lose an important feature of the optical potential, since they are all local. The off-shell behaviour of t can be taken into account by means of a simple model assumption:

$$t\left((\vec{p}+\vec{q})/2;\ (\vec{p}-\vec{q})/2\right) = g(p)\ t_E(q) \quad , \quad g(o) = 1 \quad ; \tag{7}$$

the optical potential assumes then the form:

$$\mathcal{V}_c(\vec{r},\vec{r}\,') = H\ (\vec{r}-\vec{r}\,')\ U\ (\vec{r},\vec{r}\,') \quad ,$$

with

$$H\ (x) = (2\pi)^{-3} \int d\ \vec{p}\ \exp(i\vec{p}.\vec{x})\ g(p) \quad ,$$

$$U\ (\vec{r},\vec{r}\,') = (A-1) \int d\vec{q}\ \exp(i\vec{q}.(\vec{r}+\vec{r}\,')/2)\ t_E(q)\ F\ (\vec{q};\vec{r}-\vec{r}\,') \quad .$$

If the $\vec{r}-\vec{r}\,'$ dependence in the form factor is disregarded, that is if the Fermi motion is neglected, the Frahn-Lemmer potential is obtained:

$$\mathcal{V}_c(\vec{r},\vec{r}\,') = H\ (\vec{r}-\vec{r}\,')\ U\ (\vec{r}+\vec{r}\,')/2) \quad .$$

Assuming

$$g(p) = \exp\ (-\beta^2 p^2/4) \quad , \tag{8}$$

then H(x) has the Perey-Buck form:

$$H\ (x) = (\pi\beta^2)^{-3/2}\ \exp(-x^2/\beta^2) \quad ; \tag{9}$$

β is so the non-locality range. The potential U(y) is given by a folding rule:

$$U(y) = (A-1) \int d\vec{q}\ \exp(i\vec{q}.\vec{y})\ t_E(q)\ F(q) \quad . \tag{10}$$

The potential of Eq. (5) has many other terms, beyond the simple diagonal one of Eq. (6). In order to study them it is convenient to express the t-matrix as a matrix element in spin space of an operator $M_E^j\ (\vec{k},\vec{k}')$, which, according to the Wolfenstein parametrization,[10] is the sum of five terms; however, in the case of the interaction of nucleons with spherical core nuclei, only two of them survive, so that:

$$\mathcal{V}_E(\vec{r},\vec{r}') = \mathcal{V}_c(\vec{r},\vec{r}') + \mathcal{V}_s(\vec{r},\vec{r}') \quad ,$$

where $\mathcal{V}_c$ is given by eq. (6) and:

$$\mathcal{V}_s(\vec{r},\vec{r}') = (2\pi)^3\ (A-1) \int d\vec{p}\ d\vec{q}\ \exp\ (i\vec{p}.(\vec{r}-\vec{r}'))\exp(i\vec{q}.(\vec{r}+\vec{r}')/2)$$

$$.F\ (\vec{q};\vec{r}-\vec{r}')\ t^s((\vec{p}+\vec{q})/2,(\vec{p}-\vec{q})/2)\ (\vec{p}x\vec{q})\cdot\vec{\sigma} \quad .$$

Now, in analogy to Eqs.(7),(8) the factorization for t^s is assumed:

$$t^s((\vec{p}+\vec{q})/2,\ (\vec{p}-\vec{q})/2) = -i\ g_s(p)\ t_E^s(q)\ r_\pi^{\ 2} \quad ,$$

with

$$g_s(p) = \exp\ (-\beta_s^2 p^2/4) \quad .$$

Then (neglecting a term of the order β_s^2 in the spin part), the final expression of the potential is:

$$\mathcal{V}_E(\vec{r},\vec{r}') = H(x)\ U(y) + \vec{\sigma}\ .\vec{L}\ H_s(x)\ U_s(y) \quad ,$$

(11)

$$x = |\vec{r}-\vec{r}'| \quad , \quad y = |\vec{r}+\vec{r}'|/2 \quad ,$$

where H(x), U(y) are given by Eqs. (9) and (10) respectively and

$$H_s(x) = (\pi\beta_s^{\ 2})^{-3/2}\ \exp\ (-x^2/\beta_s^{\ 2}) \quad ,$$

$$U_s(y) = r_\pi^{\ 2}/y\ (d/dt)\ U_\sigma(y) \quad ,$$

$$U_\sigma(y) = (A-1) \int d\vec{q} \;\; e^{i\vec{q}\cdot\vec{y}} F(q) \; t_E{}^s(q) \quad ;$$

r_π is the pion Compton wavelength and has been extracted for dimensional convenience.

The non-local potential (11) is in general energy dependent, but in the low energy region the non-local optical potential can be assumed to have no intrinsic energy dependence.[9,11] So the parameter E in Eq. (11) can be suppressed.

In order to perform phenomenological analyses, the form of U(y) and $U_s(y)$ must be specified.

Since F(q) has a forward peak, the following expansions for the complex t-matrices can be introduced:

$$t(q) = t_o + t_1 q^2 \quad ,$$

$$t^s(q) = t_o^s + t_1^s q^2 \quad .$$

The first term gives rise to a volume potential in the central part and to a surface one in the spin-orbit interaction. The second term of course originates a surface central potential, which, at not too high energies, is much more important for the imaginary part than for the volume one. The spin-orbit potential is in principle complex too, but its imaginary part can be neglected, at least up to the π -meson threshold.[12,8] As a rule, the volume interactions are assumed of the Woods-Saxon type, the surface ones as derivatives of Woods-Saxon. Thus, the model potential is taken as:

$$(12) \qquad \mathcal{V}(\vec{r},\vec{r}\,') = H(x)\; U(y) + \vec{\sigma}\cdot\vec{L}\; H_s(x)\; U_N{}^s(y) \quad ,$$

where:

$$U\;(y) = U_N(y) + iW_N(y) \quad ,$$

$$U_N(y) = -V_N\, f_N(y) \quad ,$$

$$W_N(y) = -4W_N f_N(y)\,(1-f_N(y)) \quad ,$$

$$U_N^s(y) = -V_{Ns} r_\pi^2 \,/(ra_N) f_N(y)\,(1-f_N(y)) \quad ,$$

with

$$f_N(y) = (1+\exp\,((y-R_N)/a_N))^{-1} \quad .$$

A consequence of the preceding considerations is that the geometrical parameters of all the three potentials are the same. The free parameters are then seven energy-independent quantities: the depths V_N, W_N, V_{Ns}, the geometrical parameters R_N, a_N and the non-locality ranges β , β_s .

4. The phenomenological analysis

For practical purposes it is convenient to introduce a local energy-dependent potential which is equivalent to the non-local one given in Eq. (12).

Following the technique used by Fiedeldey,[13,14] generalized to the case in which also the spin-orbit potential is non-local, the equivalent potential can be written as:

$$U_L(r) = U_L^o(r) + V_1(r)\,\vec{L}^2 + \vec{\sigma}.\vec{L}\,V_s(r) \quad ,$$

where

$$U_L^o(r) = U_R(r) + i\,W(r) \quad , \alpha = \beta^2\, m/2\hbar^2 \quad ,$$

$$V_1(r) = \alpha_s\,(V_s(r))^2 \quad , \alpha_s = \beta_s^2\, m/2\hbar^2 \quad ,$$

$$V_s(r) = U_L^{so}(r)/(1-\alpha U_L^o(r)),$$

m is the reduced nucleon mass.

The potentials $U_L^o(r)$ and $U_L^{so}(r)$ are given by a Perey-Buck transformation:

$$U_L^o(r) = U(r) \exp(-\alpha(E-U_L^o(r))) \quad , \tag{13}$$

$$U_L^{so}(r) = U_N^s(r) \exp(-\alpha_s(E-U_L^o(r))) \quad . \tag{14}$$

As a consequence of the transformation, the central potential acquires a small l-dependent surface term. The transformation for the central real and imaginary parts can be extracted under the assumptions that U_N^s is real and $\alpha W \cong \alpha_s W \cong 0$, since W is usually rather small. Consequently, the transformation for the real central potential does not depend on W:

$$U_R(r) = U_N(r) \exp(-\alpha(E-U_R(r))) \quad , \tag{15}$$

$$W(r) = W_N(r) \exp(-\alpha(E-U_R(r))) \quad , \tag{16}$$

and the local spin-orbit interaction is real.

The transformation written above could be used to calculate numerically the equivalent local potential. It is however preferable to assume that the Woods-Saxon form is preserved by the transformation,[15] that is:

$$U_R(r) = -V_L f(r,R_V,a_V) \quad ,$$

$$W(r) = -4Wf(r,R_I,a_I)(1-f(r,R_I,a_I)) \quad ,$$

$$V_s(r) = -2V_s f(r,R_s,a_s)(1-f(r,R_s,a_s))/r \quad .$$

The smoothness parameters can be assumed to be the same for all potentials, as the numerical fits seem to be rather insensitive to it, provided that it is not far from standard values. In the present analysis:

$$a_V = a_I = a_s = a_N = 0.57 \text{ fm} \quad .$$

Then the transformations (14),(15),(16) supply a relation between the non-local parameters and the local ones:

(17) $$V_L(E) = V_L(o) + b\,E + c\,E^2 \quad ,$$

(18) $$b = -\alpha V_L(o)/(1+\alpha V_L(o)) \quad ,$$

(19) $$c = b^2/(2\,V_L(o)\,(1+\alpha V_L(o))) \quad ,$$

(20) $$R_V(E) = R_N + a_N \ln\,(2\exp\,(\alpha V_L(E)/2) - 1) \quad ,$$
$$W\,(E) = W_N \exp\,(-\alpha\,(E+V_L(E)/2)) \quad ,$$

$$R_I(E) = R_N + a_N \ln((1 - f_{NI})/f_{NI}) \quad ,$$

(21) $$V_s(E) = V_s(o)\exp\,(-\alpha_s\,(E+V_L(E)/2-V_L(o)/2))$$
$$\cdot\,(1+\alpha V_L(o)/2)/(1+\alpha V_L(E)/2) \quad ,$$

$$R_s(E) = R_N + a_N \ln\,((1-f_{Ns})(1-f)/(f_{Ns}f)) \quad .$$

E is defined as the centre of mass energy, minus the Coulomb shift energy between the nucleon and the core nucleus:

$$E = E_{CM} - V_{Coul} \quad .$$

The quantities f_{NI}, f_{Ns}, f are definite functions of $V_L(E)$, α and α_s.[6] We can now consider as free parameters not all the non-local ones, but also some selected local quantities, namely;

(22) $$V_L(o),\ b,\ r_N = R_V(o)/(A-1)^{1/3},\ W(o),\ V_s(o),\ \alpha_s \quad .$$

The non-local parameters are completely determined by these quantities. In fact:

$$V_N = V_L(o)\exp\,(-b/(1 + b)) \quad ,$$

$$\alpha = -\,b/(V_L(o)\,(1 + b)) \quad ,$$

$$R_N = R_V(o) - a_N \ln(2 \exp(\alpha V_L(o)/2)-1) \quad ,$$

$$W_N = W(o) \exp(\alpha V(o)/2) \quad ,$$

$$V_{Ns} = V_s(o)(1 + \alpha V_L(o)/2) \exp(\alpha_s V_L(o)/2) \quad .$$

The parameter search is as follows (see Fig. 1). Some starting values for the parameters are chosen. The energy dependence of the radii is so completely determined. The potential depths V_L, V_s are fitted to the single-particle charge distributions for various values of r_N,

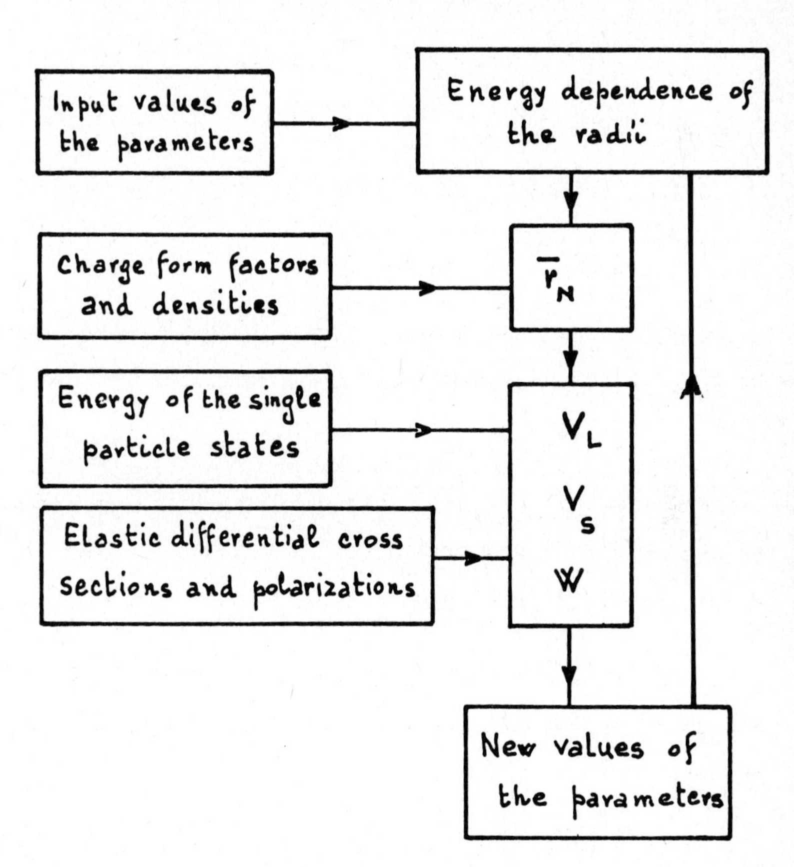

Fig. 1. Block-diagram for the parameter search.

until a satisfactory agreement is obtained. Experimental data on charge distributions are only fragmentary for the nuclei just above the closed shells. The fitting of the charge distributions is performed then for the "core" nuclei ^{12}C, ^{16}O, ^{28}Si, ^{32}S, ^{40}Ca (Figs. 2,3). In order to calculate the charge density or the monopole form factor,the knowledge of the occupation number is essential. The shell model values give satisfactory results only for the p-shell nuclei. From Si to Ca, the experiments support the hypothesis of a progressive filling of the $2\,S_{1/2}$ shell, so the occupation numbers for the outer shells are determined in order to reproduce the form of the charge density. The charge distribution is determined by the wave function corresponding to the non-local potential; therefore the results of the equivalent local potential have been corrected for the Perey effect. The resulting occupation numbers are reported in Table I. Since for each nucleus a diffe-

TABLE I

Radial parameters r_N and occupation numbers for the N = Z nuclei

Nucleus	r_N (fm)	$N_{1d_{5/2}}$	$N_{2s_{1/2}}$	$N_{1d_{3/2}}$	$N_{1f_{7/2}}$
^{12}C	1.20	-	-	-	-
^{16}O	1.29	-	-	-	-
^{28}Si	1.24	5.2	0.8	-	-
^{32}S	1.25	6.0	1.6	0.4	-
^{40}Ca	1.26	6.0	1.9	4.0	0.1

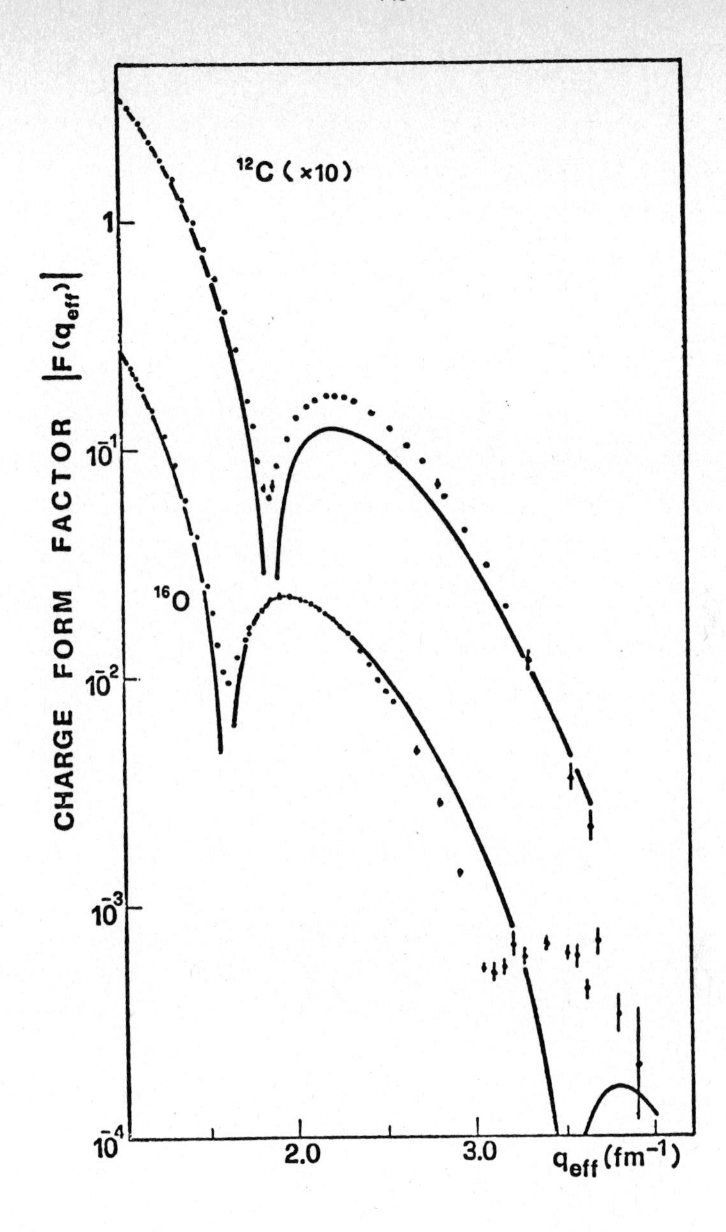

Fig. 2. Elastic monopole form factors of ^{12}C and ^{16}O. The experimental data are from Ref. 18, the curves are the independent particle model fits.

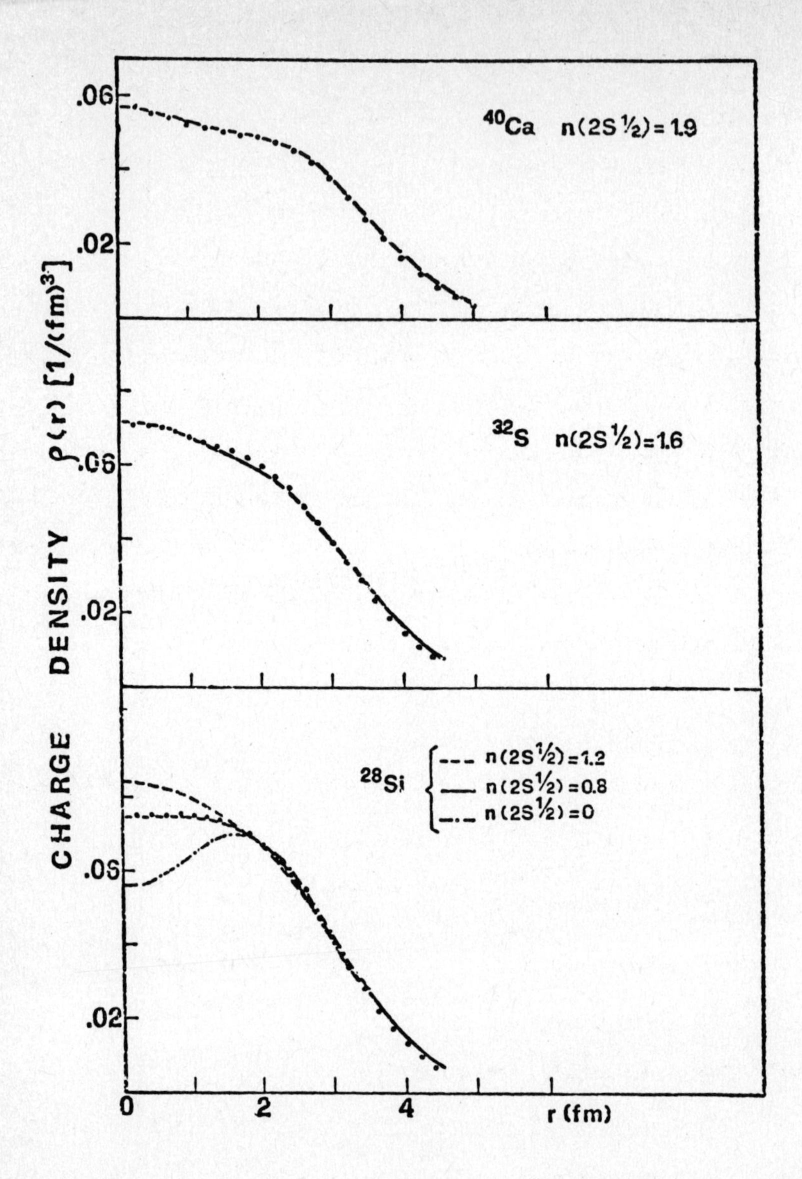

Fig. 3. Charge distributions in ^{40}Ca, ^{32}S and ^{28}Si. The experimental data are from Ref. 19, the curves are the independent particle model fits. For ^{28}Si the dependence on the 2 $S_{\frac{1}{2}}$ occupation number is shown.

rent r_N is obtained, a mean value $\bar{r}_N$ is assumed; the potential depths V_L and V_s are fitted to the energies of bound and quasi-bound single-particle states for nuclei having a N = Z core plus an extra nucleon. In the positive energy region, the three depths V_L, V_s, W are fitted to the elastic differential cross section and polarizations, the energy dependence of the various radii being completely determined. A plot is made of V_L, V_s, W versus the energy and these quantities are fitted to the model equations (Eqs.(17) to (21)) in order to supply new values of the parameters (22). The procedure is repeated from the beginning (Fig. 1). A few iterations were sufficient to obtain a satisfactory convergence. The final values of the potential depths V_L, V_s, W are given in Figs. 4 and 5. Typical fits of the differential elastic cross section and polarization are reported in Figs. 6 and 7.

5. Results and conclusions

The local parameters, determined by means of the procedure previously described, are:

$$V_L(o) = 50.5 \text{ MeV} \quad , \quad b = -0.36 \quad ,$$

$$V_s(o) = 6.0 \text{ MeV} \quad , \quad \bar{r}_N = 1.25 \text{ fm} \quad ,$$

$$W(o) = 17.6 \text{ MeV} \quad .$$

The corresponding non-local ones are:

$$V_N = 88.6 \text{ MeV} \quad , \quad \alpha = 1.11 \times 10^{-2} (\text{MeV})^{-1} \quad ,$$

$$V_{Ns} = 9.2 \text{ MeV} \quad , \quad \alpha_s = 7.10^{-3} (\text{MeV})^{-1} \quad ,$$

$$W_N = 23.3 \text{ MeV} \quad , \quad R_N = (1.25(A-1)^{1/3} - 0.285) \text{ fm} \quad ,$$

$$a_N = 0.57 \text{ fm} \quad .$$

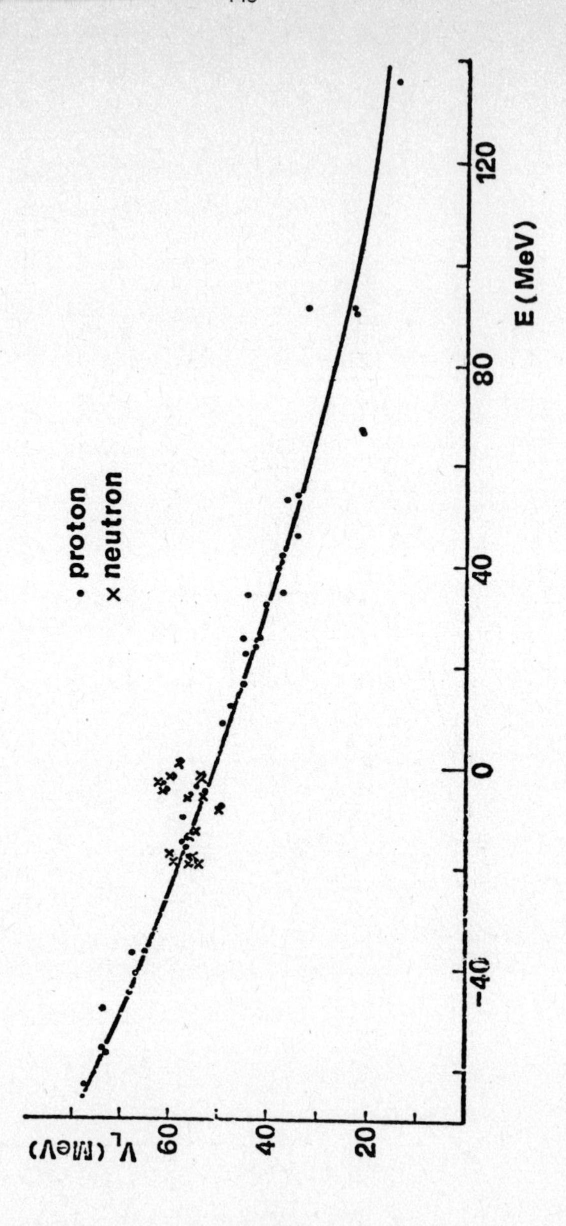

Fig. 4. Real central potential depth as a function of $E=E_{CM}-V_{Coul}$. The points are the phenomenological values, the curve is the model local equivalent potential.

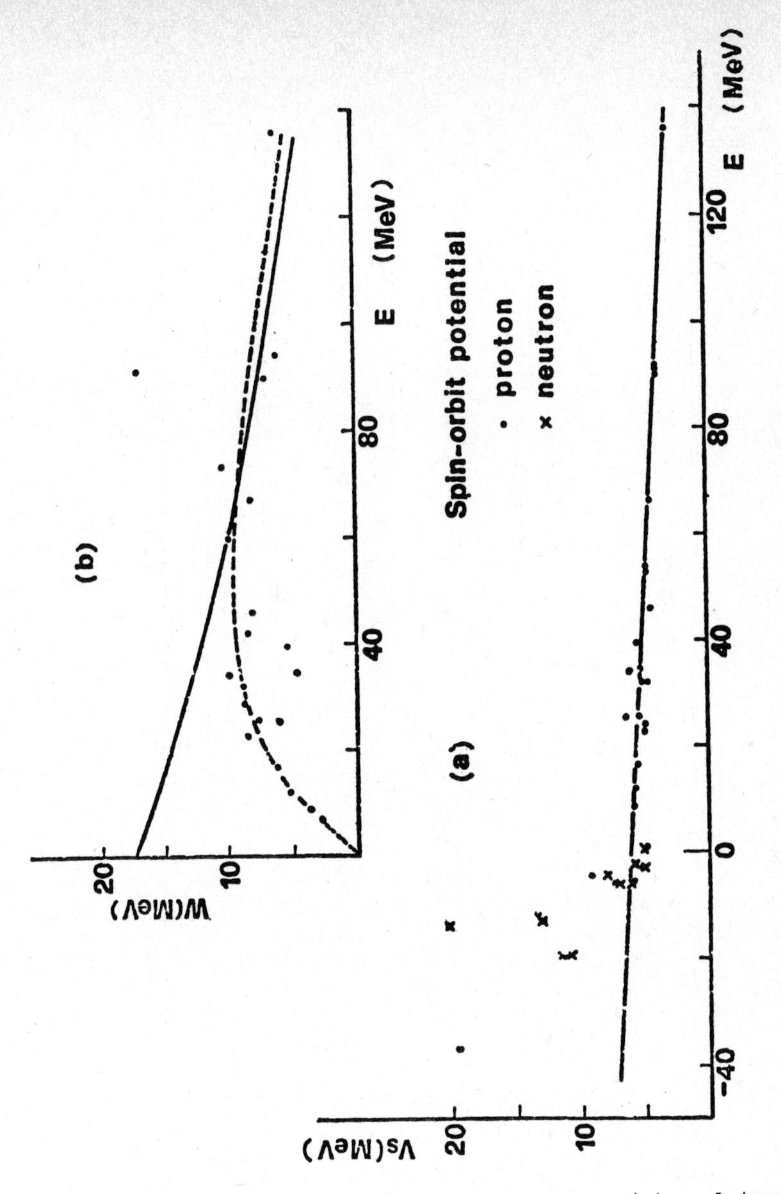

Fig. 5. The same as in Fig. 4 for the spin-orbit (a) and imaginary (b) potential depths. The broken curve in (b) shows the effect of a dynamical energy dependence on the local imaginary potential.

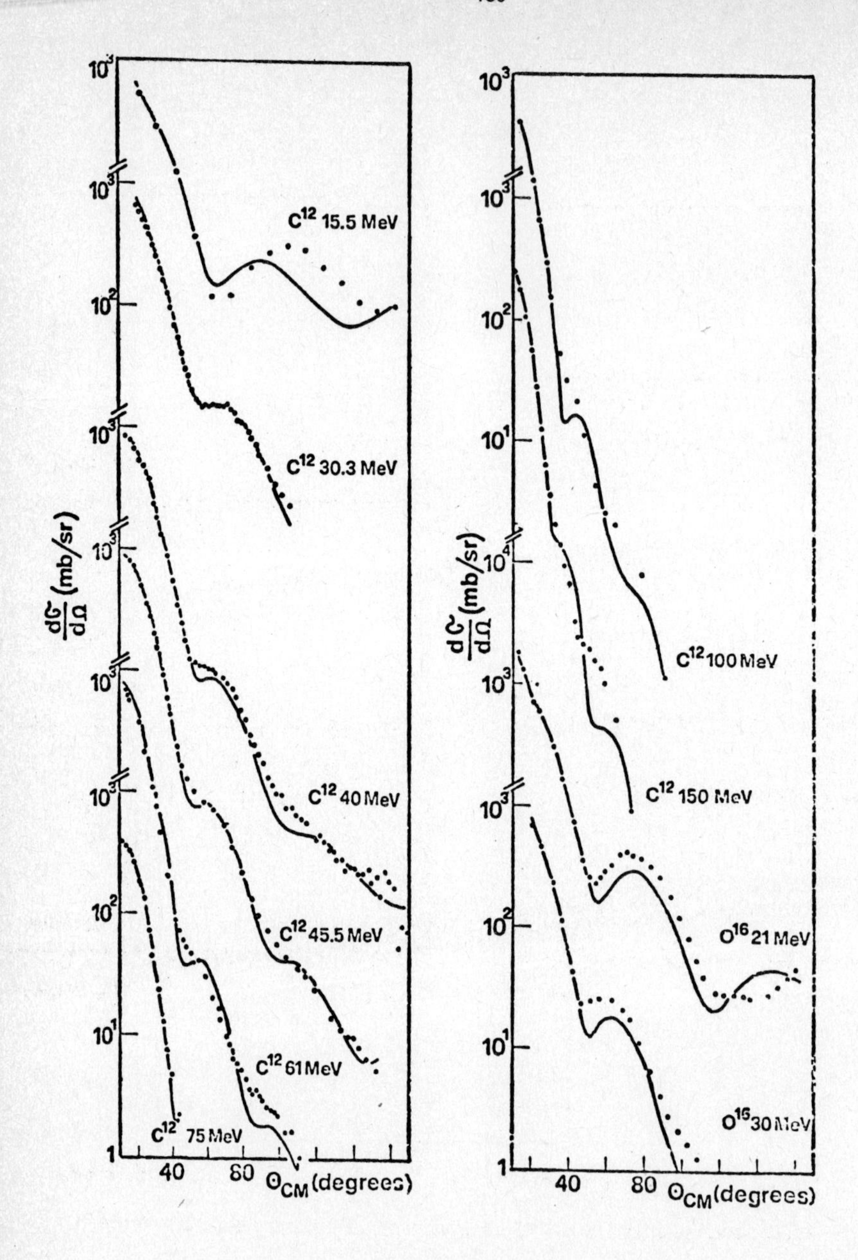

Fig.6. Proton elastic differential cross sections on ^{12}C and ^{16}O at different laboratory energies. The curves are our optical model fit. The experimental data are from Refs.20 to 28.

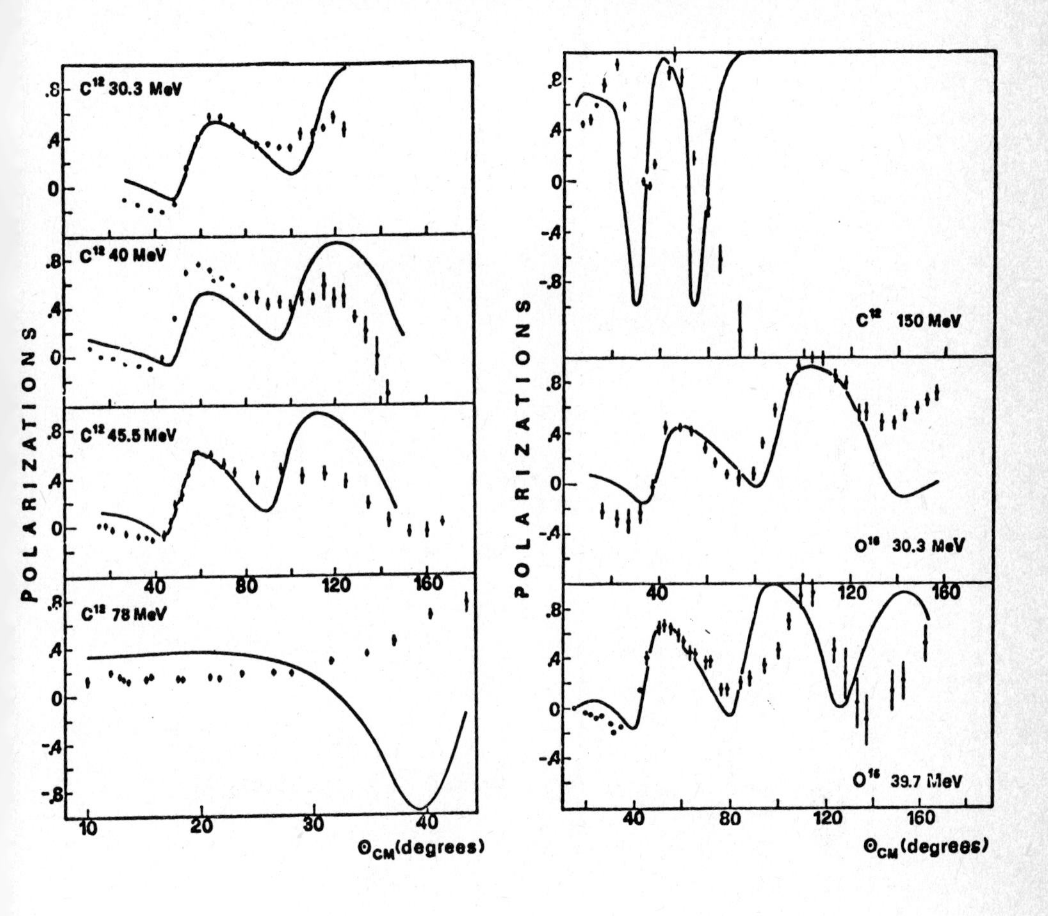

Fig. 7. Proton elastic polarization on ^{12}C and ^{16}O at different laboratory energies. The curves are our optical model fits. The experimental data are from Refs. 29, 22, 23, 25, 28 and 30.

As it can be seen from Figs.4 and 5, a non-local energy independent optical potential is a useful concept for describing bound and scattering single-particle states in an energy range of the order of 200 MeV and for a wide class of nuclei. In particular, the energy behaviour of the real central depths is in agreement with previous phenomenological analyses and also with recent theoretical calculations;[16] there is a fluctuation around E = 0, but it can be explained in terms of a non-negligible intrinsic energy dependence of the non-local potential.[17]

As for the imaginary part, the hypothesis of energy independence of the non-local potential is consistent with the analysis only at energies greater than ~60 MeV. At lower energies, the equivalent local potential and the phenomenological one are quite different.

It is therefore necessary to introduce an intrinsic energy dependence of the kind:

$$W_N(E) = W_N(1 - \exp(-\Gamma E)) \quad .$$

If

$$W_N = 29 \text{ MeV} \quad , \qquad \Gamma = 2.44 . 10^{-2} (\text{MeV})^{-1} \quad ,$$

the equivalent local potential reproduces the phenomenological values quite well (see Fig. 5b, broken line).

References

1) F.D.Becchetti jr. and G.Greenlees, Phys. Rev. 182, 1190 (1969).

2) B.A.Watson, P.P.Singh and R.E. Segel, Phys. Rev. 182, 977 (1969).

3) G.Passatore, Introductory talk of this Conference.

4) J.P.Jeukenne, A.Lejeune and C.Mahaux, "Theoretical investigation of the optical-model potential", presented at this Conference.

5) G.W.Greenlees, G.J.Pyle and Y.C.Tang,Phys.Rev. 171, 1115 (1968).

6) M.M.Giannini and G.Ricco, "An energy independent non-local potential model for bound and scattering states", to be published.

7) A.L.Fetter and K.M.Watson, "The optical model" in "Advances in Theoretical Physics" (K.A.Brückner Ed.) Vol.I, Academic Press, New York, 1965.

8) S.Gamba, G.Ricco and G.Rottigni, Nucl. Phys. A213, 383 (1973).

9) G.Passatore, Nucl. Phys. A110, 91 (1968).

10) L.Wolfenstein, Ann. Rev. Nucl. Sci, 6, 43 (1956).

11) F.Perey and B.Buck, Nucl. Phys. 32, 353 (1962).

12) L.R.B. Elton and A.Swift, Nucl. Phys. A94, 52 (1967).

13) H.Fiedeldey, Nucl. Phys. 77, 149 (1966).

14) F.Capuzzi, "Equivalent potentials in the description of scattering processes", presented at this Conference.

15) A.Gersten, Nucl. Phys. A96, 288 (1967).

16) J.P.Jeukenne, A.Lejeune and C.Mahaux, "Many-body theory of the optical-model potential", presented at this Conference.

17) C.Mahaux, Seminar given in Genova.

18) I.Sick and J.S.Mc Carthy, Nucl. Phys. A150, 631 (1970).

19) G.C.Li, I.Sick and M.R.Yearian, Phys. Rev. C9, 1861 (1974).

20) Y.Nagahara, Journ. Phys. Soc. Japan 16, 133 (1961).

21) B.K.Ridley and J.F.Turner, Nucl. Phys. 58, 497 (1964).

22) L.N.Blumberg, E.E.Gross, A.van der Wonde, A.Zucker and R.H.Bassel, Phys. Rev. 147, 812 (1966).

23) G.R.Satchler, Nucl. Phys. A100, 497 (1967).

24) C.B.Fulmer, J.B.Ball, A.Scott and M.Whiten, Phys. Rev. 181, 1565 (1969).

25) C.Rolland, B.Geoffrion, N.Marty, M.Morlet, B.Tatisheff and A.Willis, Nucl. Phys. 80, 625 (1966).

26) S.K.Mark, P.M.Portner and R.B.Moore, Can. J. Phys. 44, 2961 (1966)

27) N.Baron, R.F.Leonard and D.A.Lind, Phys. Rev. 180, 978 (1969).

28) P.D.Greaves, V.Hnidzo, J.Lowe and O.Karban, Nucl. Phys. A179, 1 (1972).

29) R.M.Craig, J.C.Dore, G.W.Greenlees, J.Lowe and D.L.Watson, Nucl. Phys. 83, 493 (1966).

30) R.N.Body, J.C.Lombardi, R.Mohan, R.Arking and A.B.Robbins, Nucl. Phys. A182, 571 (1972).

THE SYSTEMATICS OF THE $1f_{7/2}$ NEUTRON SINGLE-PARTICLE ENERGIES

F.MALAGUTI

Istituto di Fisica dell'Università, Bologna, Italy

Abstract: The centroid energies of neutron particle and hole states in the $1f_{7/2}$ shell are combined to give single-particle energies using the definition of Baranger. It is shown that the latter can be better expressed as eigenvalues of a local Saxon-Woods potential.

One of the simplest nuclear models describes the nucleus as an assembly of independent particles each one moving in a common potential whose eigenvalues are called "single-particle energies".
In the simplest version of the theory, the single-particle states are fully occupied up to the Fermi level, and completely empty thereafter. It is then possible to measure the energy of an occupied state by measuring the energy necessary to remove a nucleon from it, and the energy of an empty state by measuring the energy released putting a nucleon in it. The first is a pickup experiment and gives a "hole energy", the second is a stripping experiment and gives a "particle energy".

If the states under consideration are really full or empty, it is possible to identify the hole and particle energies with the single particle energies, and to represent them as eigenvalues of a one body, local Saxon-Woods potential with depth depending on A and on the nuclear asymmetry parameter.[1,2]

If the state considered is only partially filled, things get more complicated. Both the corresponding hole and particle energies can be measured, and they are found to differ, as do the potentials corresponding to the hole states in regions of A where the shell is filling and where it is full.[2]

So the question arises of how to define the single-particle energy experimentally if it has to maintain the simple meaning described at the beginning. Baranger has suggested[3] to define it as an average between particle and hole energies, using as weigths the corresponding spectroscopic factors.

In Table I are collected some $1f_{7/2}$ hole and particle energies BE_i (i=1,3) together with their total spectroscopic factors S_i (i=1,3) from one-neutron transfer reactions. The hole state contains two fragments depending on the isospin of the daughter nucleus.

TABLE I

Nucleus	N	Z	Holes $T_<$		Holes $T_>$		Particles				
			BE_1	S_1	BE_2	S_2	BE_3	S_3	E_{exp}	E_{theor}	ΔE
^{36}Ar	18	18		0.	18.46	0.63	6.78	6.87	7.76	7.77	-0.01
^{40}Ca	20	20		0.	18.44	0.41	8.36	6.87	8.93	9.35	-0.42
^{42}Ca	22	20	12.85	1.87	18.60	0.28	7.75	5.49	9.40	9.43	-0.03
^{43}Ca	23	20	10.29	2.46	>17.68	0.10 a)	8.36	5.46	>9.07	9.48	-0.41
^{44}Ca	24	20	11.40	3.53	19.90	0.07	7.41	3.37	9.55	9.52	0.03
^{46}Ca	26	20	10.40	4.95	21.71 b)	0.04 b)	7.28	2.00	9.57	9.61	-0.04
^{48}Ca	28	20	10.02	7.03	19.40 a)	0.035		0.	10.06	9.71	0.35
^{48}Ti	26	22	12.25	5.31	19.01	0.46	7.56	2.49	11.21	10.94	0.27
^{50}Cr	26	24	13.63	4.37	17.70	0.85	8.89	2.73	12.44	12.20	0.24
^{52}Cr	28	24	12.56	7.20	18.67	0.79	5.84	0.90	12.42	12.21	0.21
^{54}Cr	30	24	11.96	5.57	20.37	0.56	5.28	0.90	11.77	12.23	-0.46
^{54}Fe	28	26	14.47	4.47	17.62	1.60	7.28	2.01	13.31	13.36	-0.05
^{56}Fe	30	26	12.93	7.09	18.98	0.92		0. c)	13.63	13.35	0.28
^{58}Ni	30	28	15.07	3.94	17.78	2.79	6.39	1.21	14.70	14.40	0.30
^{90}Zr	50	40	16.78	7.70	25.08	0.90		0.	17.65	17.90	-0.25

a) From Ca isotopes systematics.

b) From energy and spectroscopic factor of analogue state + Coulomb shift + sum rules.[4]

c) From extreme single-particle model and sum rules.

These energies have been averaged to give the single-particle energies

$$(1) \qquad E_{exp} = \sum_{i=1}^{3} S_i BE_i \Big/ \sum_{i=1}^{3} S_i \quad ,$$

and the result is again displayed in Table I (+). All energies are in MeV. To see whether the Baranger energies defined by (1) or the hole (or particle) energies can be better reproduced by a one-body potential the data have been fitted to the expression

$$(2) \qquad V(r) = V f_1(r) + V_s \left(\frac{\hbar}{m_\pi c}\right)^2 \frac{1}{r} \frac{df_2(r)}{dr} \vec{L}.\vec{\sigma} \quad ,$$

where the form factors $f_{1,2}(r) = (1+\exp((r-R_{1,2})/a_{1,2}))^{-1}$, with $R_1 = 1.10A^{1/3} + 0.75$, $R_2 = 1.10A^{1/3}$, $a_1 = 0.52$, $a_2 = 0.65$ fm, and $V_s = 7$ MeV. The potential depth is written

$$(3) \qquad V = V_o - 4V_1 \vec{t}.\vec{T} / A + \gamma A \ .$$

The experimental hole and particle energies have been first analysed separately to give the optimum values for V_o, V_1 and γ using the method described in Refs. 1,2. The result is displayed in the first three lines of Table II, showing the optimum values for V_o, V_1, γ and the root-mean-square deviation σ_E between calculated and experimental energies.

TABLE II

	V_o	V_1	γ	σ_E
Hole energies $T_<$	55.54	34.40	-0.113	0.65
Hole energies $T_>$	65.43	17.58	-0.214	1.81
Particle energies	66.81	13.10	-0.519	0.75
Baranger energies	49.61	17.89	-0.047	0.27

(+) E_{exp} corresponds to $-\varepsilon$ in Ref. 3.

If the particle and hole energies are averaged using (1), the resulting combined centroids may also be represented by optimum V_o, V_1 and γ . This calculation may be simplified by analysing all the data together with a potential having an averaged isospin term, and this is justified because these terms are small.

This average isospin potential is for neutron states

$$V_1 = \left(S_1\left(-\frac{N-Z-1}{A-1}\right) + S_2\left(\frac{N-Z+3}{A-1}\right) + S_3\left(-\frac{N-Z}{A}\right)\right) V_1 \Big/ \sum_{i=1}^{3} S_i \quad ,$$

where S_1 , S_2 , S_3 can be expressed in terms of the average neutron and proton occupation numbers in the shell, $\langle\nu\rangle$ and $\langle\pi\rangle$, using the sum rules of French and MacFarlane.[4)]

With the approximation $\frac{1}{A-1} \simeq \frac{1}{A}$, one gets the potential depth

$$V = V_o + \frac{V_1}{A}\left(-N+Z+\frac{\langle\nu\rangle+2\langle\pi\rangle}{2j+1}\right) + \gamma A. \tag{4}$$

In calculating E from V, one has to take into account the fact that A to be used in solving the Schrödinger equation is different for particle and for hole states. It is reasonably approximate to solve the wave equation only once by using an "effective value" for A of the whole system (particle + "core").

$$A_{eff} = A + 1 - \frac{\langle\nu\rangle}{2j+1} \quad . \tag{5}$$

The occupation numbers $\langle\nu\rangle$ and $\langle\pi\rangle$ have been evaluated in Eqs. (4) and (5) according to the extreme single-particle model. The potential depths V corresponding to E_{exp} and Eqs. (2) and (5) have been fitted to Eq. (4) using a least-squares procedure, and the results are displayed in the fourth line of Table II. The corresponding calculated energies E_{theor} are collected in Table I together with the differences $\Delta E = E_{exp} - E_{theor}$.

The quality of the fit, shown by σ_E, is clearly better for the Baranger energies than for the hole and particle energies, and the observed deviations in the former case are comparable with the experimental uncertainties.

This indicates that in the region of A where a shell is filling, such a definition of single-particle energies can be better connected with the simple idea of independent motion in a one-body potential. This definition has been subjected to some criticism concerning far-away fragments that could make it meaningless,[5] but the results of the present analysis show that such effects could be not very important.

References

1) D.J.Millener and P.E.Hodgson, Nucl. Phys. A209, 59 (1973).

2) F.Malaguti and P.E.Hodgson, Nucl. Phys. A215, 243 (1973).

3) M.Baranger, Nucl. Phys. A149, 225 (1970).

4) J.B.French and M.H.MacFarlane, Nucl. Phys. 26, 168 (1961).

5) C.A.Engelbrecht and H.A.Weidenmüller, Nucl. Phys. A184 385 (1972).

SMALL ANGLE ELASTIC SCATTERING OF POLARIZED PROTONS[†]

G. Bendiscioli, E. Lodi Rizzini, A. Rotondi, M.L. Stanga, A. Venaglioni

Istituto di Fisica Nucleare,Università di Pavia,Pavia

Istituto Nazionale di Fisica Nucleare,Sezione di Pavia

Abstract. Preliminary results on asymmetry and cross section between 0° and 7° in the elastic scattering of 36.2 MeV polarized protons off C, Al, Fe, Ni targets are given.

1. Introduction

We report on the first results of an experiment now in progress at the cyclotron in Milan on elastic scattering at small angles ($\Theta_{LAB} < 7^{\circ}$) of 40 MeV polarized protons. There are several experimental results on asymmetry and differential cross section at small angles and with energies above 90 MeV. At lower energies there are a few results on the cross section and only one on the asymmetry relative to the argon nucleus.[1)]

It is wellknown that elastic scattering at small angles is produced mainly by the interaction of the protons with the nuclear surface, where the attractive nuclear potential becomes of the same order of magnitude as the repulsive Coulomb potential and the spin-orbit potential is maximum.

The elastic scattering cross section for polarized protons scattered by spin-zero nuclei may be written in the form

$$\sigma(\theta,\phi) = \sigma(\theta) \quad \left(1+A(\theta)\vec{P}x\vec{n}\right) \quad ,$$

where θ is the scattering angle, ϕ is the azimuthal angle, $\vec{P}$ is the

---

† presented by E. Lodi Rizzini

polarization of the incoming beam and $\vec{n}=(\vec{K}_i \times \vec{K}_f)/|\vec{K}_i \times \vec{K}_f|$, where $\vec{K}_i$ e $\vec{K}_f$ are the initial and final proton momenta; $\sigma(\theta)$ is the differential cross-section averaged over ϕ and $A(\theta)$ is the asymmetry parameter.

Semi-qualitative arguments[2] show that at small angles the cross-section and the asymmetry do not vary monotonically with angles, but have an oscillatory behaviour, which is attributed to the interference between Coulomb and nuclear scattering.

Among other things, elastic scattering measurements at small angles are of interest:

a) in a more precise determination of optical model parameters, particularly of the spin-orbit parameters;[3]
b) in the removal of the ambiguity in the phase shift analysis of scattering of protons on light nuclei;[4]
c) in clarifying the differences between the proton and neutron distributions in the nuclei.[5]

It is to be noted that at small angles measurements of cross-section and asymmetry permit, in principle, to determine completely the nuclear scattering amplitude, without the necessity of performing a triple scattering experiment.

2. Experimental set-up

In Fig. 1 it is shown a schematic view of the experimental set-up, which at present accepts a 42 MeV proton beam extracted from the Milan Cyclotron.[6]

By scattering at about 57,5° in the horizontal plane on a 90 mg/cm^2 thick carbon target (T_1) and by means of an analyzing magnet and a collimator system, a beam of partially polarized protons ($P \simeq 0.8$, $E \simeq 36.2$ MeV) is selected. The polarized protons are scattered by the target

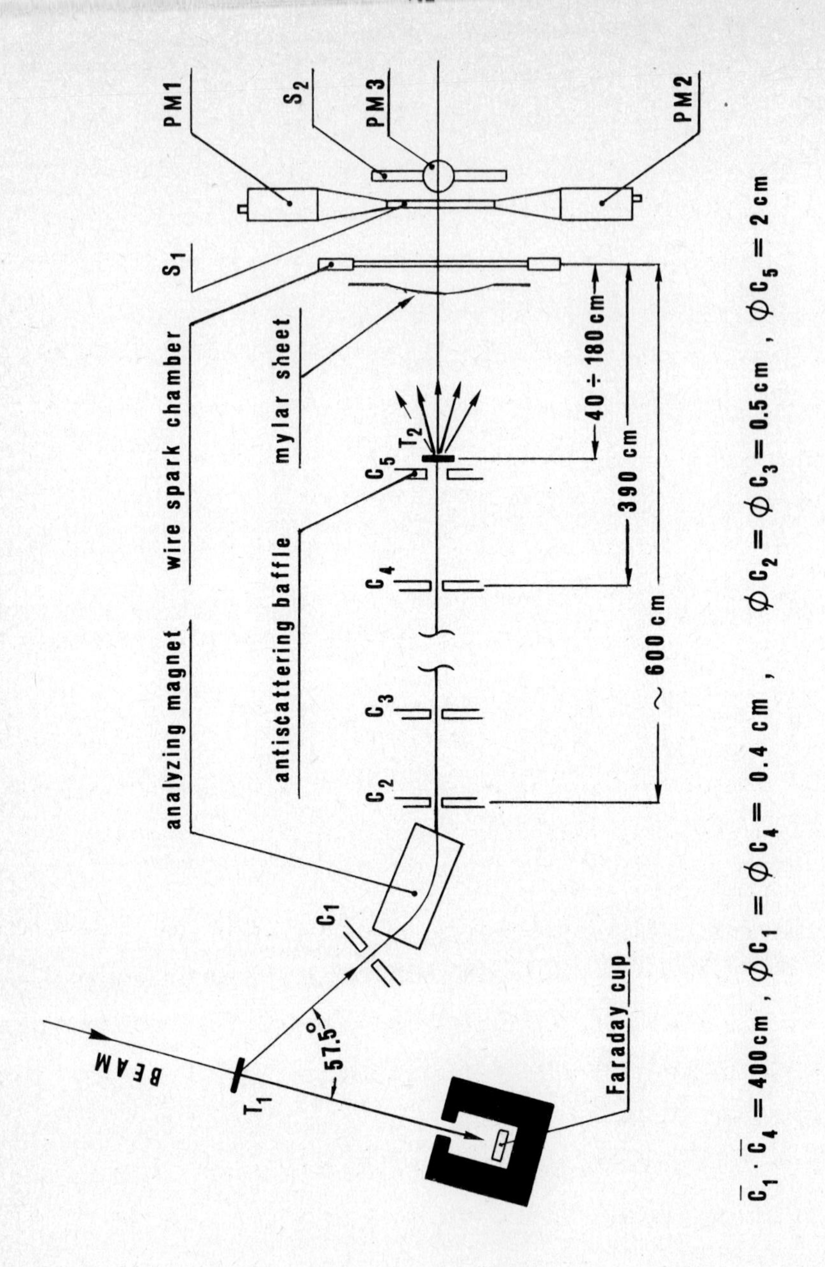

Fig. 1. Schematic view of the experimental set-up.

T_2 and then detected by the magnetostrictive wire spark chamber (30x30 cm^2, wire spacing=0.1 cm) and the scintillator S_1.

The apparatus is based on the use of a magnetostrictive readout wire spark chamber triggered by a plastic detector. It permits to measure the trajectories of protons scattered into a cone whose angular acceptance depends on the dimension of the chamber and on the distance between this and the target. This distance is variable to allow measurements with different angular resolution. By choosing properly this distance and the diameter of the collimators, it is possible to obtain an uncertainty on the scattering angle of the order of a few primes. Working without target, the spark chamber permits to know with high accuracy the distribution of the protons in the incident beam, this information is very important to avoid errors due to geometrical uncertainties in the data reduction.

A signal generated in the scintillator starts an operation sequence, which begins with the firing of the chamber and ends with the storing on magnetic tape of the coordinates of a spark; the storing is performed by a Laben 70 on-line computer.

The magnetostrictive read-out wire spark chamber is a conventional one; it has two planes of ortogonal wires and gives two pairs of magnetostrictive signals. The time interval between the signals of a pair defines the value of one of the coordinates of the point where a proton crossed the chamber.

Some difficulties of small angle experiments depend on the fact that the differential cross section varies with the scattering angle θ as θ^{-4}; thus the measurements are very sensible to the uncertainties on θ depending both on lack of precision in the geometry of the experimental set-up and on the finite size of the beam and of the detectors.

These difficulties have been overcome in a satisfactory way.

A further important cause of uncertainty is the Coulomb multiple scattering on the targets, whose effects increase with decreasing energy, as the cross section varies with energy as E^{-2}.

In the data reduction the greatest difficulties arise from the finite beam cross section and from the multiple scattering in the target; unfortunately we could not use the apparatus in the best working conditions because of the low polarized beam intensity ($10 \div 20$ protons per sec); to obtain high statistics in a short time, a very narrow bear or very thin targets cannot be used.

3. Data Reduction

The first goal of our experiment is the determination of the distribution $N(\theta)$ of the protons scattered by the targets and the mean asymmetry $\langle A(\theta)\rangle$. The later one is defined in the following way.

For the sake of simplicity we neglect momentarily the finite cross section of the beam and the multiple scattering on the target.
Let us consider the reference system of Fig. 2, associated with the wire spark chamber.

Let us consider equal intervals $\Delta_i\theta=\theta_{i+1}-\theta_i$; we indicate with $N_L(\Delta_i\Omega)$ the number of the protons deviated on the left (i.e. into the solid angle defined by $\Delta_i\theta$ and $-\pi/2<\phi<\pi/2$) and with $N_R(\Delta_i\Omega)$ the number of the protons deviated on the right (i.e. into the solid angle defined by $\Delta_i\theta$ and $\pi/2<\phi<3\pi/2$) moreover we put $N(\Delta_i\Omega)=N_L(\Delta_i\Omega) + N_R(\Delta_i\Omega)$.

We define the measurable quantity

$$A_m(\Delta_i\Omega) = \frac{N_L(\Delta_i\Omega)-N_R(\Delta_i\Omega)}{N(\Delta_i\Omega)} .$$

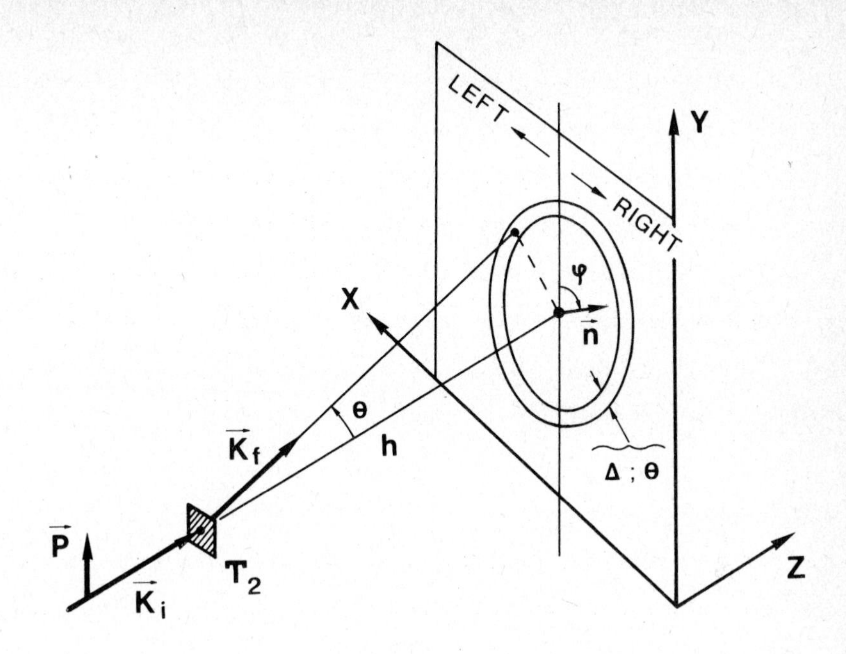

Fig. 2. Reference system.

By the formula $\sigma(\theta\phi)=\sigma(\theta)\left(1+A(\theta)\vec{P}x\vec{n}\right)$ we have

$$A_m\ (\Delta_i\Omega)\ =\ P\langle\cos\phi\rangle\langle A(\theta)\rangle\quad,$$

where $\langle\cos\phi\rangle=2/\pi$ is the value of $\cos\phi$ averaged over $-\pi/2<\phi<\pi/2$ and

$$\langle A(\theta)\rangle\ =\ \frac{\int_{\Delta_i\theta}\sigma(\theta)A(\theta)d\theta}{\int_{\Delta_i\theta}\sigma(\theta)d\theta}$$

is the mean value of $A(\theta)$ over $\Delta_i\theta$. Thus

$$\langle A(\theta)\rangle\ =\frac{\pi}{2}\,\frac{1}{P}\ A_m(\theta)\ . \tag{1}$$

The statistical error on $A_m(\Delta_i\Omega)$ is

$$\sigma_i = \sqrt{\frac{1-A_m^2(\Delta_i\Omega)}{N(\Delta_i\Omega)}} \quad .$$

Results of the same kind are obtained if intervals $\Delta\phi$ different from π are considered. For instance, in the next section we will consider the events whose scattering plane forms an angle smaller than 45° with the horizontal plane and the remainder separately. For the two sets of events we have

$$\text{(2)} \qquad \langle A(\Theta)\rangle = \frac{\pi}{2\sqrt{2}} \; \frac{1}{P} \; A_{<45}(\Theta) \quad ,$$

$$\text{(3)} \qquad \langle A(\Theta)\rangle = \frac{\pi}{2(2-\sqrt{2})} \; \frac{1}{P} \; A_{>45}(\Theta) \quad .$$

In effect the experimental distributions of the protons are affected by the finite beam cross section and by the multiple scattering. They may be thought as obtained by means of a convolution among the scattering cross section σ, the Molière function f_M describing the Coulomb multiple scattering and the function describing the proton distribution in the incoming beam g_B.

Following Cormack [7] we have:

$$f_{exp} = f_M * g_B + Nt(f_M * \sigma_n) * g_B \quad .$$

N is the number of the scattering centers per cm^3 and t the target thickness; $\sigma_n = \sigma - \sigma_R$ where σ_R is the Rutherford cross-section: * means a convolution. We recall that, because of the convolutions, f_{exp} appear flatter and larger than σ.

The same considerations are true for the asymmetry.

The quantity g_B is known from the experiment because, while measuring, target-in runs alternated with target-out runs.

However in the results reported on in the next section the corrections for the finite beam cross section and the multiple scattering are limited to the Coulomb scattering.They consist in substituting,after an appropriate normalization, to the experimental event distribution $N_{exp}(\theta)$ the quantity

$$N(\theta) = N_{exp}(\theta)-(N_{MB}(\theta)-N_R(\theta)) \quad ,$$

where N_{MB} is the number of events prevised by the convolution $f_M * g_B$ and $N_R(\theta)$ is the corresponding number due only to single Rutherford scattering. The origin of the scattering angle Θ is on the axis normal to the chamber and passing the beam center, whose coordinates are determined by g_B.

$N_{MB}-N_R$ is a negative quantity near 0° and a positive one elsewhere; it goes slowly to zero while angle increases. The normalization requires that the theoretical distribution $N_{MB}(\theta)$ coincides at the smallest angles ($\lesssim .5^\circ$) with $N_{exp}(\theta)$.

Possible effects on the asymmetry due to the beam cross-section were analyzed by simulating the experiment with a computer using the observed distribution of the proton in the beam and the Molière function.

The data are to be corrected also for the background, which may be evaluated either by analyzing the events in the region out the cone of the scattered protons or by considering the events out the incoming beam in the target-out runs.

4. Results

We have performed measurements with four natural targets (C, Al, Fe, Ni; see Table I) between 0 and 7 degrees. The distance between the target and the spark chamber was 100 cm for C and 60 cm for the other nuclei.

TABLE I

Target	Z	A (percent natural abundance).	Thickness mg/cm^2	r.m.s. multiple scattering angle
C	6	12 (99)	84	$.56^{\circ}$
		...		
Al	13	27 (100)	84	$.80^{\circ}$
Fe	26	54 (5.8) 56 (91.7)	87	1.15°
		...		
Ni	28	58 (67.8)	99	1.26°
		60 (26.2)		

Half of the data were collected with the detector system (spark chamber and scintillator) upright and half with the detector system upside-down, i.e. rotated on the incident beam direction of 180°. In this way, by comparing the two distributions $A_m(\Delta_i\Omega)$ obtained with the upright and upside-down system, we can put in evidence possible asymmetries due to the apparatus.

The four targets were exposed to the beam alternatively; every target was changed after about 10^5 events.
Our results are summarized in the fig. 3, 4, 5 and 6.

The observed distributions are completely corrected for the background and partially for the finite beam cross section and multiple scattering, as stated in the previous section.

At the smallest angles there is a lack of events, for which N/N_R

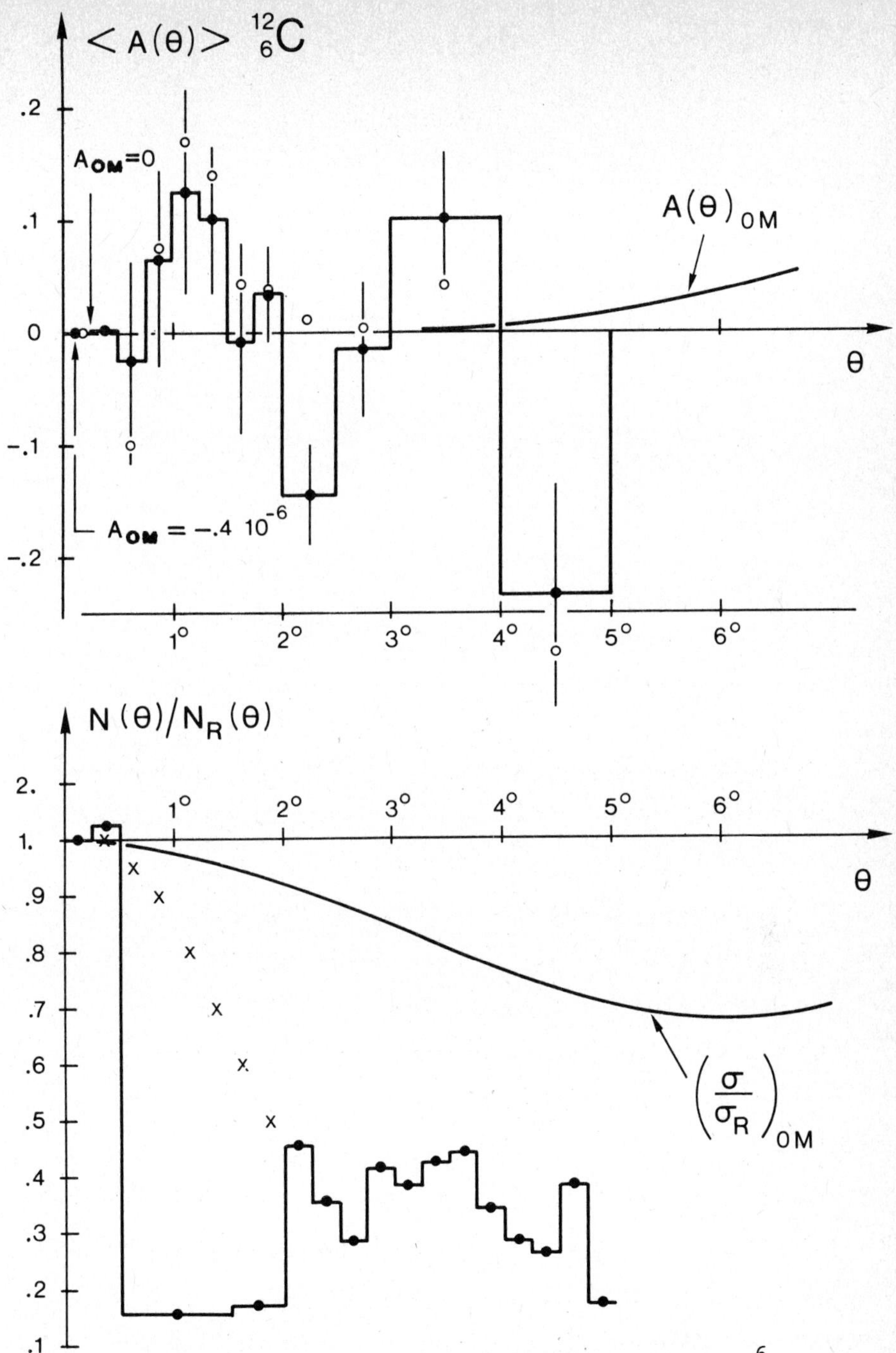

Fig. 3. Energy = 36.2 MeV. Number of events = 2.1×10^6; OM = optical model previsions. x Interpolated points. o$\langle A(\theta)\rangle_{<45^\circ}$

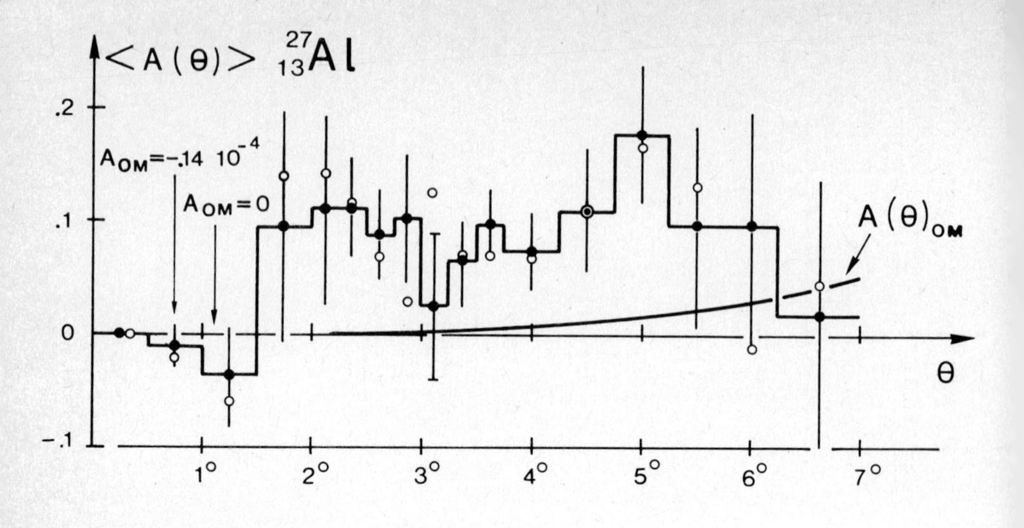

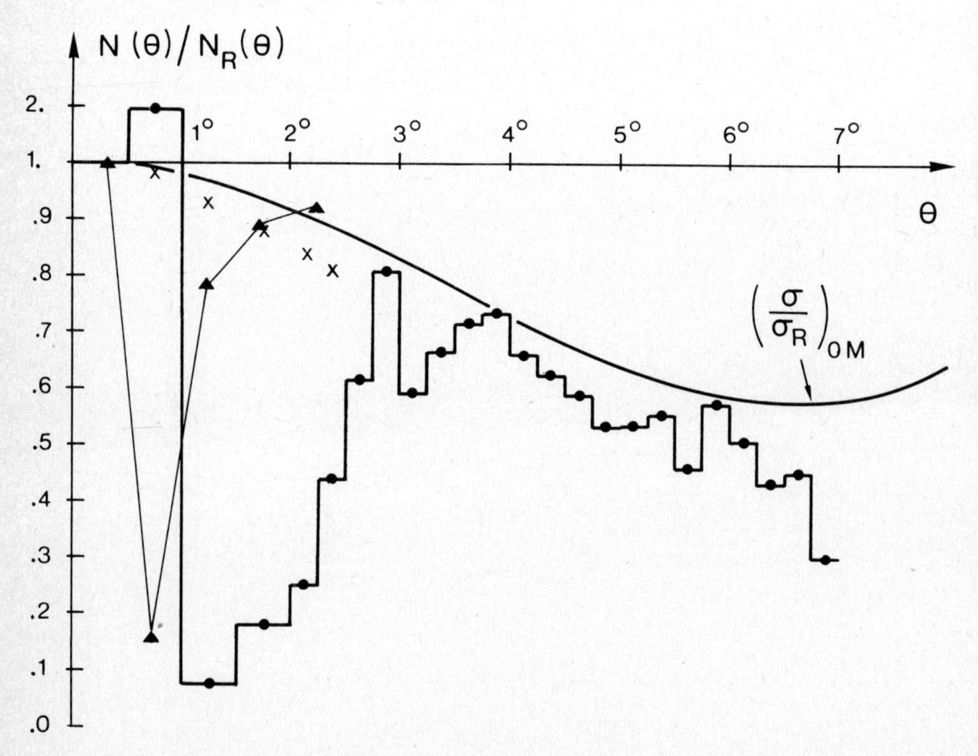

Fig. 4. Energy = 36.2 MeV. Number of events = 1.9×10^6. OM = optical model previsions. x Interpolated points. Δ Thin targets events. o$\langle A(\theta)\rangle_{<45^\circ}$

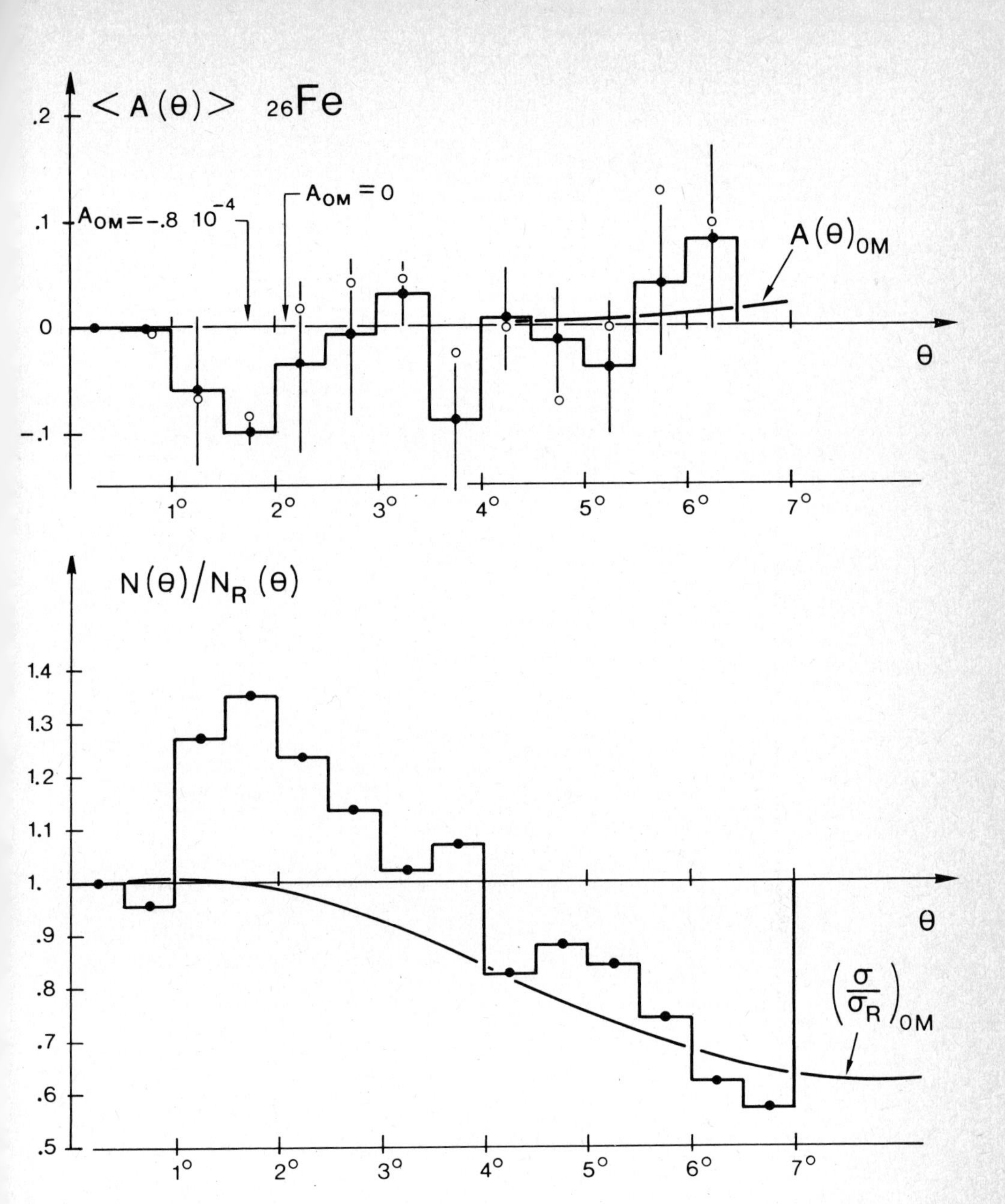

Fig. 5. Energy = 36.2 MeV. Number of events = 1.5×10^6. OM = optical model previsions. x Interpolated points. o$\langle A(\theta)\rangle_{<45}$.

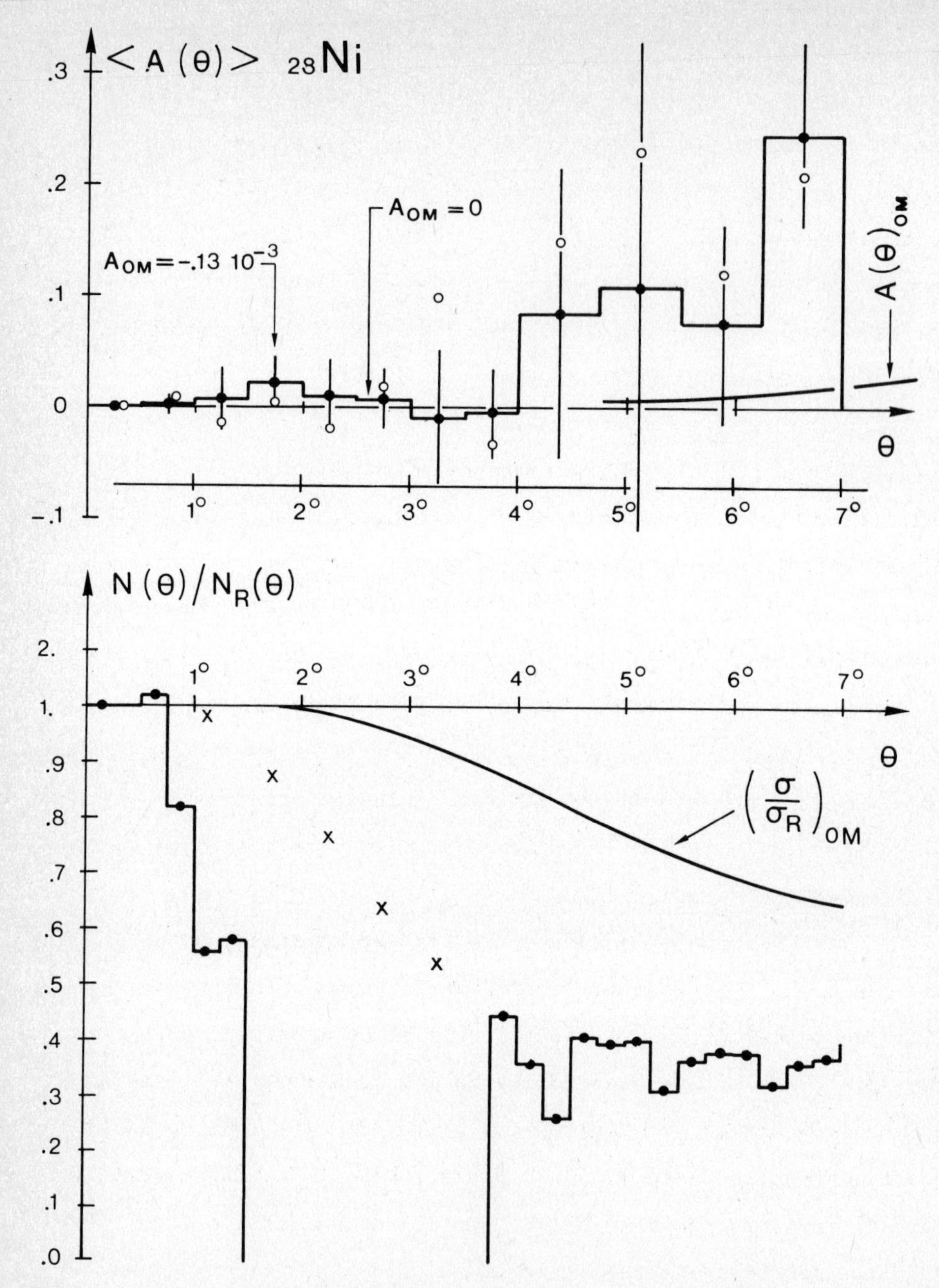

Fig. 6. Energy = 36.2 MeV. Number of events = 1.2×10^6; OM = optical model previsions. x Interpolated points. o$\langle A(\theta)\rangle_{<45}$.

appears too much smaller than 1; even if oscillations of the cross-section are not to be excluded, the lack of events is due mainly to the inadequacy of the corrections. In this region, in the evaluation of the asymmetry we have introduced a rough approximation (see the interpolated points in the figures), which likely overestimates the number of events. The correctness of this procedure is confirmed by the behaviour of N/N_R obtained with a thin Al target (14 mg/cm^2) at great distance from the wire chamber (200 cm) (see Fig. 4).

In evaluating the asymmetry, to put in evidence possible effects due to the apparatus and the presence of random errors other than the statistical ones, the events whose scattering half plane forms with the horizontal plane an angle smaller than 45° and the remainder were considered separately. We calculated the asymmetries of the two sets of events (see Eqs. (2) and (3)), their mean (see Eq. (1)) and the standard deviation of the mean. This last results to coincide with the statistical error at the larger angles, while is greater at the smaller ones. The asymmetries of the pairs of sets agree very well with the mean with a reduced χ^2 equal to 1.075, 0.987, 1.13, 0.88 for C, Al, Fe and Ni respectively.

The most remarkable thing in the results is the discrepancy with the phenomenological optical model previsions.[8] Preliminary calculations show that it is not reduced neither by sensible variations of the spin orbit parameters, nor by taking in account the interaction between the magnetic moment of the proton and the Coulomb field of the nucleus. Moreover we point out that a behaviour of the asymmetry different from that prevised by the optical model was put in evidence previously (see Fig. 7) in an experiment of scattering on argon nuclei with a diffusion cloud chamber.[1]

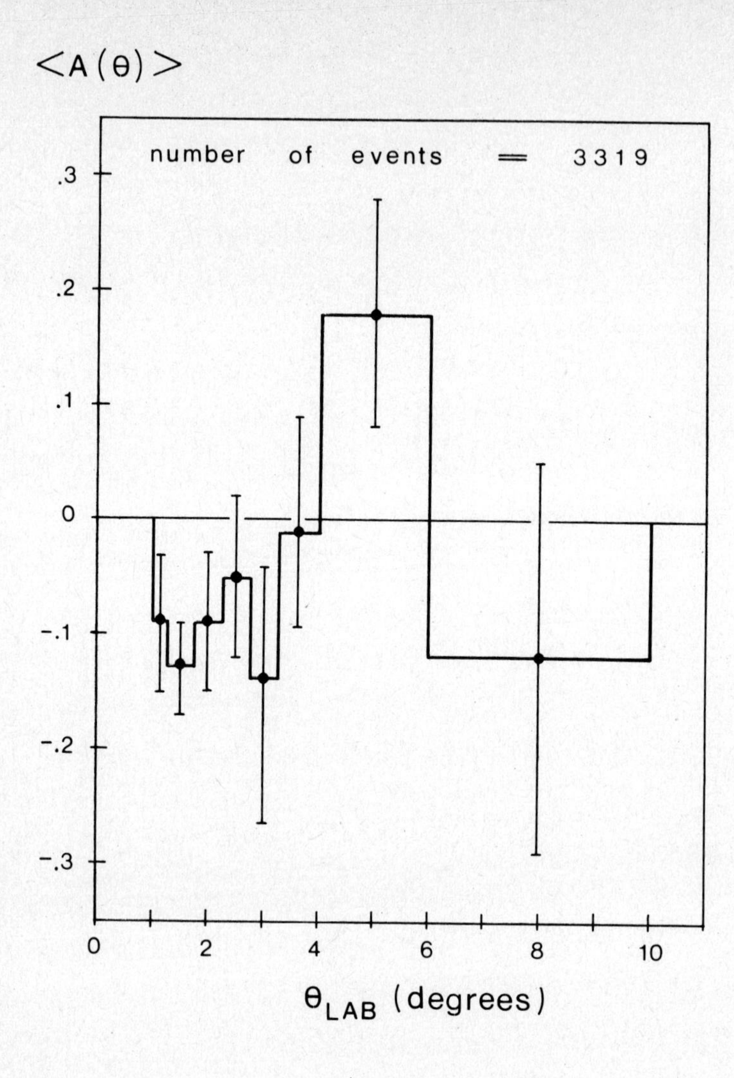

Fig. 7. Diffusion cloud chamber experiment on argon nuclei at 39.5 MeV.

We retain to be premature to give any interpretation of the results; it is our intention to minimize the residual multiple scattering and beam size effects and to perform other measurements with the sake of enriching the statistics at the larger angles and increasing the angular resolution at the smaller ones.

Acknowledgement

Thanks are due to Mr. C.Marciano of the I.N.F.N. Section of Pavia and to the staff of Cyclotron of Milan for their valuable collaboration.

References

1) G.Bendiscioli, A.Gigli Berzolari, E.Lodi Rizzini, Nuovo Cimento 27A, 27 (1975).

2) G.Bendiscioli, A.Gigli Berzolari, E.Lodi Rizzini, Lettere al Nuovo Cimento 6, 687 (1973)

3) L.J.B.Goldfarb, G.W.Greenlees and M.B.Hooper, Phys.Rev. 144, 829 (1966).

4) C.C.Giamati and R.M.Thaler, Nucl. Phys. 59, 159 (1964); S.A.Harbison, R.J.Griffiths, N.M.Steward, A.R.M.Johnston and G.T.A.Squier, Nucl. Phys. 150A, 570 (1970); G.E.Thompson, M.B.Epstein and T.Sawada, Nucl. Phys. 142A, 571 (1970).

5) G.W.Greenlees, G.J.Pyle and Y.C.Tang, Phys. Rev. 171, 1115 (1968); G.W.Greenlees, C.H.Poppe, J.A.Stevers, D.L.Watson, Phys. Rev. C3, 1063 (1970).

6) G.Bendiscioli, E. Lodi Rizzini, C.Marciano, A.Rotondi, A.Venaglioni, Nucl. Instr. and Meth. 124, 397 (1975).

7) A.M.Cormack, Nucl. Phys. 52, 286 (1964).

8) F.D.Becchetti Jr. and G.W.Greenlees, Phys. Rev. 182, 1190 (1969);
B.A.Watson, P.P.Singh and R.E.Segal, Phys. Rev. 156, 1207 (1967).

DISPERSION RELATION ANALYSES OF THE ENERGY DEPENDENCE OF THE OPTICAL POTENTIAL

G. PASSATORE

Istituto di Scienze Fisiche dell'Università di Genova
Istituto Nazionale di Fisica Nucleare - Sezione di Genova

Abstract. The various attempts made to separate the spurious energy dependence of the empirical optical potential from the dynamical one by using a dispersion relation for the theoretical optical potential are reviewed in the light of a recent calculation of the mass operator in the Brueckner-Hartree-Fock approximation. The non-relativistic and relativistic treatments are compared and their implications are pointed out. A comparison between the calculations based on the dispersion relation and those based on the nucleon-nucleon scattering amplitude is also made.

1. Introduction

The energy dependence of the empirical local potential

$$\mathcal{V}_L(E) = -V_L(E) - i\, W_L(E) \tag{1}$$

has two different origins. One lies in the own energy dependence of the theoretical non-local potential itself and gives rise to the so called "dynamical energy dependence" of the empirical potential. The second is due to the fact that the empirical local potential is local while the theoretical potential is non-local: the additional arising energy dependence is called spurious energy dependence.

In this paper I shall summarize a few attempts that have been made to separate the spurious from the intrinsic energy dependence of the empirical local potential by using the dispersion relation which formally holds for the theoretical optical potential.[1,2]

The most of these works date from some years ago[3-7]: in this paper they will be discussed in the light of the recent theoretical optical potential calculation made by Jeukenne, Lejeune and Mahaux in the BHF approximation [8] (which has been illustrated in a previous paper in this meeting) and their results will be also compared with the implications of the formulation of the theoretical optical potential in terms of nucleon-nucleon scattering amplitude[9,10]. Moreover a very recent work[11] on the dispersion relation analysis will be considered in this survey.

Section 2 contains a survey of the phenomenological situation. In Section 3 the basic formulae used in the dispersion-relation analyses are shown. In Section 4 it is discussed whether the empirical data satisfy the dispersion relation. Section 5 describes the various attempts made in order to separate the spurious energy dependence from the dynamical one: Section 5.1 deals with the non-relativistic treatments, Section 5.2 with the relativistic ones. Finally, in Section 6 a comparison between the results of the dispersion-relation calculation and those based on the nucleon-nucleon scattering amplitude is made.

2. The phenomenological situation

First of all, let us give a look at the results of the phenomenological analyses for the real and imaginary potential depth $V_L(E)$ and $W_L(E)$ in the interior of a nucleus.[4] The data refer to a range of nuclei, from light nuclei such as carbon to heavy such as lead. The analyses have been made with well-depths of various type, such as square, trapezoidal, Saxon - Woods. So the collection of these data has only the aim to give general trends, but here just these properties

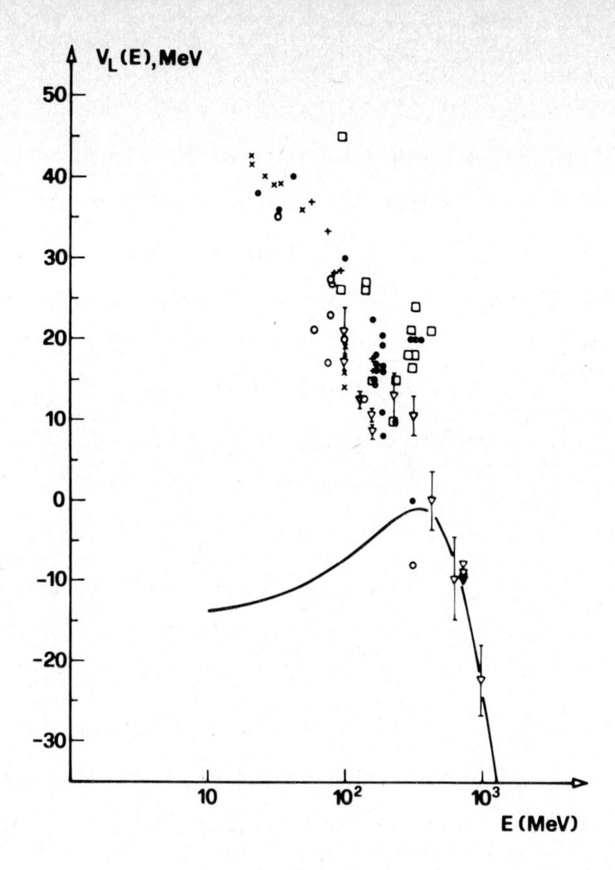

Fig.1. The real part of the local potential depth from empirical analyses:plusses:neutrons, volume absorption;crosses: neutrons,surface absorption;black dots:protons, volume absorption;circles: protons,surface absorption;squares:protons,non relativistic analyses;triangles:protons,relativistic analyses; superimposed cross and circle:average for neutrons and protons. Full line: subtracted dispersion integral obtained from the imaginary part shown in Fig. 2. (From. Ref. 4).

are now concerned. We observe that:

1) The real part of the empirical potential depth $V_L(E)$ is monotonically decreasing with the energy and it changes sign around 300 MeV (transition from attraction to repulsion). (Fig. 1)

2) The imaginary part of the potential depth $W_L(E)$ is monotonically increasing with the energy and goes asymptotically to a constant value (Fig. 2). The black triangles are obtained not by direct nucleon-nucleus scattering analyses but by using the relation:

$$W_L(E) = \frac{1}{2} \hbar v_{lab} \gamma \rho \bar{\sigma} \quad , \tag{2}$$

where $\bar{\sigma}$ is the nucleon-nucleon experimental total cross section averaged on protons and neutrons, ρ is the nuclear density, γ a factor accounting for the Pauli principle. The constant asymptotic behaviour, then, is determined by the asymptotic behaviour of the nucleon-nucleon total cross-section.

Formula (2) is a classical formula for the attenuation of a beam travelling in a homogeneous absorptive medium. It can also be derived in the approach, sketched in the introductory talk, which expresses the theoretical optical potential in terms of the nucleon-nucleon t-matrix, as we are going to discuss at the end of this paper. By using Eq. (2) implicitely one gives to the optical potential a relativistic meaning, because he takes into account all the anelastic channels due to particle production. On this point the points of view of the various authors are different. Some authors prefer to consider the optical potential in a non-relativistic context. They still maintain the formula (2) but calculate the cross-section σ by a non relativistic model for the nucleon-nucleon interaction, which leads to a vanishing imaginary forward scattering amplitude at high energy.

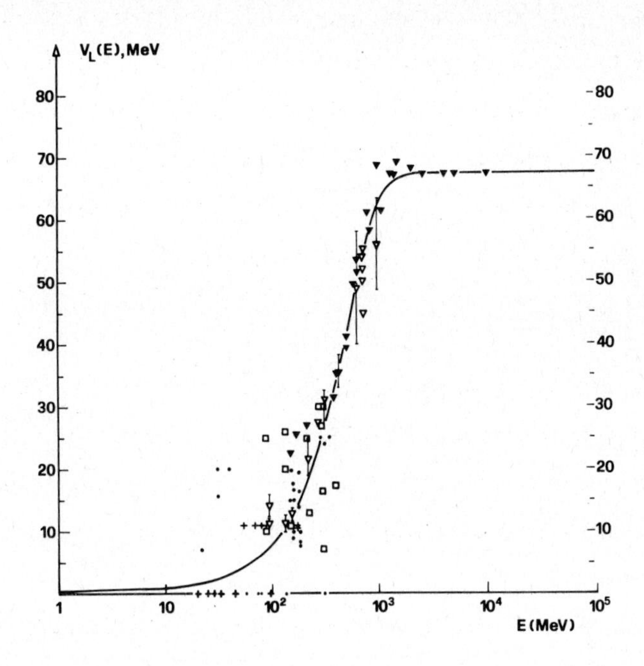

Fig.2. The imaginary part of the local potential depth from empirical analyses and from nucleon-nucleon cross-sections:plusses:neutrons;black dots: protons;squares: protons,non relativistic analyses;triangles:protons, relativistic analyses;black triangles: from total nucleon-nucleon cross section.

Full line: interpolation curve.

(From Ref. 4)

3. The basic formulae

The basic formulae are the following:

- Structure of the theoretical optical potential in the coordinate representation:

$$\mathcal{V}(E)\Psi(\vec{r}) = \mathcal{V}_1\Psi(\vec{r})+\int \mathcal{V}_2(|\vec{r}-\vec{r}\,'|)\Psi(\vec{r}\,')d\vec{r}\,'+\int \mathcal{U}(E;|\vec{r}-\vec{r}\,'|)\Psi(\vec{r}\,')d\vec{r}\,' , \tag{3}$$

- Local equivalent potential:

$$\mathcal{V}_L(E) = \mathcal{V}_1 + \mathcal{J}(E,k^2) \quad , \tag{4}$$

$$\mathcal{J}(E;k^2) = \int\left(\mathcal{V}_2(s)+ \mathcal{U}(E;s)\right) \exp\left(i(\tilde{\vec{k}}\cdot\vec{s})d\vec{s}\right) , \quad \vec{s} = \vec{r}-\vec{r}\,' , \tag{5}$$

$$\tilde{k}^2=\begin{cases} \dfrac{2m}{\hbar^2} \left(E- \mathcal{V}_L(E)\right) , & \text{(non relativistic)} \\ (\hbar^2c^2)^{-1} \left(E_t{}^2-m^2c^4-2E_t\mathcal{V}_L(E)\right), & \text{(relativistic, } E_t>>|\mathcal{V}_L(E)|) , \end{cases} \tag{6}$$

- Dispersion relation for the theoretical optical potential:

$$\mathrm{Re}\,\mathcal{U}(E;|\vec{r}-\vec{r}\,'|) = \frac{1}{\pi} \; P \int_0^\infty \frac{\mathrm{Im}\,\mathcal{U}(E',|\vec{r}-\vec{r}\,'|)}{E' - E} dE' \quad . \tag{7}$$

The formulae written here refer to the limit case of infinite nuclear matter. This may seem a rather crude approximation, but:

1) It may be realistic in the interior of heavy nuclei;

2) The formulae are coherent among themselves and quite simple, so one hopes they may contain the essential points and therefore lead to a good information as long as general trends are concerned.

Of course in this way the whole information about radial dependence and surface effects will be lost, but I think that such effects are far beyond from the possibility of analyses of this type.

It is true that each of the formulae given above can also be writen for finite nuclei, in an exact or approximate form, but I feel the approximations involved in their application are such that any significant information, in a matter of details, will be lost. It must be remarked, however, that if a theoretical optical potential corresponding to a scattering amplitude averaged over a suitable energy interval is considered, the dispersion relation assumes the form (7) even for finite nuclei.[5)]

4. The direct analysis of the experimental data

The first approach is to see whether the phenomenological values by themselves agree with the dispersion relation. If this were so, no spurious energy dependence would be in the empirical optical potential, i.e. the theoretical optical potential would be local.

If one had to use the dispersion relation for finite nuclei, two difficulties would appear: the presence of the poles in the dispersion relation connected with the narrow compound nucleus resonances, and the fact that the phenomenological potential concerns the energy averaged scattering amplitudes. Such difficulties can be overcome at once if one takes an average of the theoretical optical potential over the same averaging interval used in making the empirical analyses. As noted above, for such an averaged theoretical optical potential the dispersion relation in the form (7) (i.e. without poles) holds. One obtains the same result if assumes the point of view of the infinite nuclear matter: the poles of the theoretical optical potential disappear and the averaged experimental scattering amplitude may be interpreted as the scattering amplitude descrived by such a potential.

The dispersion integral can be calculated by using for the imaginary part a suitable function which interpolates the experimental data.

If one assumes a relativistic point of view, the asymptotic behaviour is given by Eq. (2) and such a function is represented by the curve in Fig. 2. A subtracted dispersion relation is required and, as a consequence, only the slope of the dispersion integral is meaningful. Such a slope, in the whole region below 400 MeV, is in complete disagreement with the data, even qualitatively (Fig. 1, full line).

A similar disagreement is found if the non-relativistic point of view is assumed (Fig. 4, dotted line). In this case for the imaginary part such a behaviour as that shown in Fig. 3 is used. Also here only

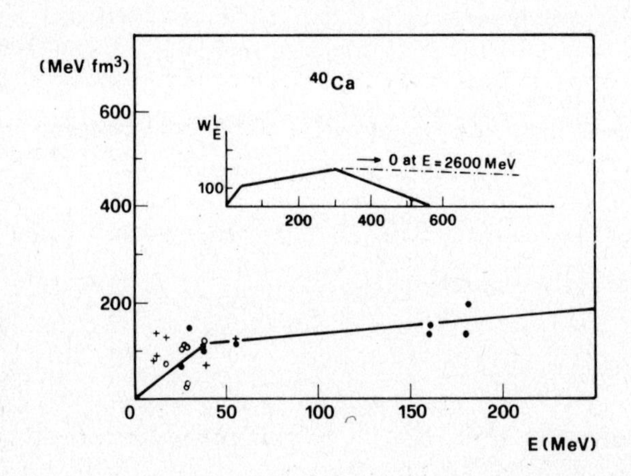

Fig.3. The imaginary part of the local empirical optical potential depth $W_L(E)$ for protons on ^{40}Ca integrated over the nuclear volume. Full points: volume absorption; open circles: volume + surface absorption; crosses: surface absorption.

Full line: the interpolation curve used in Ref.5.

Dot and dashed line: the behaviour from a non-relativistic nucleon-nucleon interaction.

(From Ref.5)

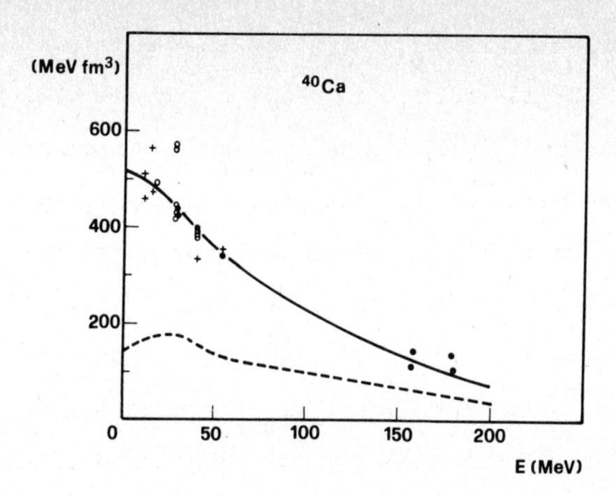

Fig.4. The real part of the local empirical optical potential depth $V_L(E)$ for protons on ^{40}Ca integrated over the nuclear volume obtained from Eq. (12) (full line).
The dotted line is the dispersion integral calculated with the imaginary part shown in Fig. 3.
(From Ref.5).

the slope of the dispersion integral is significant, because the high energy tail of the imaginary part is rather arbitrary, and this causes an additive constant to the result.

The reason that the relativistic and the non-relativistic treatments give, at low energy, similar results, is that the relativistic treatment uses a subtracted dispersion relation, and consequently the constant asymptotic behaviour of the imaginary part does not affect the results, as long as the low energy region is concerned.

The disagreement between the dispersion integral (calculated with the empirical imaginary part) and the empirical values of the real part is a very important point, because it means that the energy dependence of the empirical potential has an important spurious component at leas at these energies. Such a conclusion justifies the non local, energy-independent models introduced at low energy for the theoretical optica potential. It also agrees with the recent calculation by Jeukenne, Lejeune and Mahaux[8)] by the Brueckner-Hartree-Fock approximation. On this point we are returning later on.

5. The separation between the spurious and the dynamical energy dependence

Now the question arises how to separate the dynamical energy dependence from the spurious one and in expressing the latter in terms of the non locality of the theoretical optical potential. Of course, the empirical optical potential is treated as the equivalent local potential of the theoretical optical potential.

5.1. The non relativistic treatments

Let us first consider the non-relativistic treatments. In Ref.5 and in Ref.6 it is assumed that the non-locality is given only by the term $\mathcal{V}$ in Eq. (3) due to the identity of the nucleons and the theoretical optical potential, after averaging over a suitable energy interval, is written as:

$$\mathcal{V}(E)\Psi(\vec{r}) = \int \mathcal{V}_2(|\vec{r}-\vec{r}\,'|)\Psi(\vec{r}\,')d\vec{r}\,' + \mathcal{U}(E)\Psi(\vec{r}) \quad , \tag{8}$$

where $\mathcal{U}$ is a local complex potential. As it concerns $\mathcal{V}_2$, Gaussian or Yukawian models are assumed:

$$\mathcal{V}_2(s) = UH_\beta(s) \quad , \tag{9}$$

where

$$H_\beta(s) = (\pi\beta^2)^{-3/2} \exp(-s^2/\beta^2) \quad , \tag{10}$$

$$H_\beta(s) = (\pi\beta^2 s)^{-1} \exp(-2s/\beta) \quad . \tag{11}$$

The parameter β is a measure of the non locality. In the case of the Gaussian model, if $(\beta^2 m/2\hbar^2)W \ll 1$, by means of Eqs. (4), (5), (6) one gets:

$$V_L(E) = U\exp\left(-\frac{\beta m}{2\hbar^2}(E+V_L(E))\right) - \mathrm{Re}\,\mathcal{U}(E) \quad , \tag{12}$$

$$W_L(E)\left\{1+\frac{\beta^2 m}{2\hbar^2}\,U\exp\left[-\frac{\beta^2 m}{2\hbar^2}(E+V_L(E))\right]\right\} = -\mathrm{Im}\,\mathcal{U}(E) \quad . \tag{13}$$

Eq. (12) shows in a simple way the separation between the "spurious" energy dependence (first term at the r.h.s.) from the "dynamical" one (second term at the r.h.s.). The latter can be calculated by inserting the l.h.s. of Eq. (13) in the dispersion relation, and it results as being made up by two terms: the first one contains in the dispersion integral only the imaginary part of the empirical optical potential; the second one is more complicated but under very reasonable approximations it can be neglected.

As for the asymptotic behaviour of $W_L(E)$ is concerned, in Ref.5 a curve was drawn through the experimental points (for the cases of ^{40}Ca, ^{12}C and ^{58}Ni) which for simplicity was chosen to consist of straight-line segments and was rather arbitrarily continued to higher energies such as it goes to zero at some high energy, as suggested from the non relativistic nucleon-nucleon interaction (Fig.3). In Ref.6 for $W_L(E)$ is used the local transform of an energy independent local potential which is in good agreement with the data up to about 200 MeV and goes to zero at high energy. Both in Ref.5 and in Ref.6 the experimental points are well reproduced by Eq. (12) (Fig. 4). In Ref.5 values of the range of non-locality β ranging from 1.1 to 1.2 fm (for a Gaussian non-local model) are found; in Ref.6 the value $\beta \simeq .85$ fm for a Gaussian and 1.5 fm for a Yukawa model. In Ref.5 the dispersion relation is applied to the potential integrated over the nuclear volume and the Coulomb potential is taken into account. In Ref.6 the radial dependence of the potential is also discussed in detail with particular reference to the surface properties.

The points of view which characterize these treatments consist essentially in:

1) assuming the energy dependent part of the optical potential as local;

2) treating the optical potential as a non relativistic quantity whose imaginary part tends to zero for increasing energy.

As it concerns the first hypothesis, the recent theoretical optical potential calculation by Jeukenne, Lejeune and Mahaux[8)] in the Brueckner-Hartree-Fock approximation gives some support. In fact it is found that in the imaginary part of the empirical optical potential the dynamical energy dependence dominates on the spurious one. This is very reasonable on a physical ground, because the energy dependence of the imaginary part of the optical potential is essentially due to the opening of the anelastic channels, which is a dynamical fact. For the real part of the empirical potential the opposite situation is found. From the Brueckner-Hartree-Fock calculation the following estimates can be obtained for the average values of the derivatives of the empirical potential depths in the interval 0-100 MeV.

	Dynamical energy dependence	Spurious energy dependence
Real part	$\frac{\partial}{\partial E} V_L(E;k(E)) \sim .12$	$\frac{\partial}{\partial k} V_L(E;k(E)) \frac{dk}{dE} \sim -.42$
Imaginary part	$\frac{\partial}{\partial E} W_L(E;k(E)) \sim .2$	$\frac{\partial}{\partial k} W_L(E;k(E)) \frac{dk}{dE} \sim -.04$

Such values refer to the centre of the nucleus.

These estimates imply that the range of non locality of the imaginary part of the theoretical optical potential is less than one of the real part. But the non-locality of the real part has two origins: the term $\mathcal{V}_2$ which is due to the antisymmetrization and it is independent of the energy, and the term Re $\mathcal{U}$. The former has no counterpart in the imaginary part of the theoretical optical potential while the latter, as well as the whole imaginary part of the theoretical optical potential, comes from the spectral representation of QHQ, and is connected with the imaginary part of the theoretical optical potential by the dispersion relation. This suggests that the term responsible for the larger non-locality of the real part of the theoretical optical potential is the term $\mathcal{V}_2$, while the terms Re $\mathcal{U}$ and Im $\mathcal{U}$ would have comparable ranges of non-locality. This may justify the approximation made in the dispersion calculation above referred to by treating these terms as local ones.

In a very recent paper[11)] the non relativistic treatment is developed in a different way, without requiring any hypothesis on the non-locality of the energy dependent part. The terms $\mathcal{V}_1$ and $\mathcal{V}_2$ in the Eq. (3) are obtained by means of the Hartree-Fock approximation by using a Tabakin potential. The imaginary part of the theoretical optical potential to be inserted under the dispersion integral has been calculated in terms of the off-shell nucleon-nucleon scattering amplitude weighted with the autocorrelation function of the density fluctuations of the target. The off-shell nucleon-nucleon scattering amplitude has been obtained from the non-relativistic nucleon-nucleon interaction used to calculate the Hartree-Fock term. Such an imaginary part agrees with the experiment from about 50 MeV to few hundreds of MeV and then goes to zero (Fig. 5). In this way the real part of the theoretical optical potential, and consequently that of the empirical local poten-

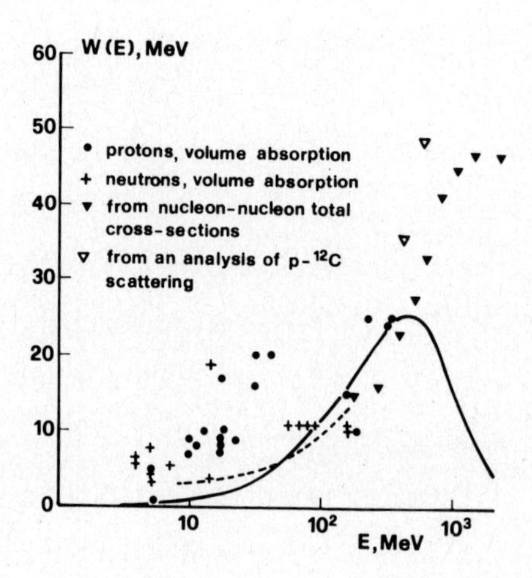

Fig. 5. The imaginary part of the equivalent local potential depth obtained from a non-local potential derived from the non-relativistic nucleon-nucleon scattering amplitude. (From Ref. 11).

-tial, can be obtained without any free parameter by using the dispersion relation in the unsubtracted form. The result is shown by the curve A in Fig. 6 which appears shifted above the experimental points by

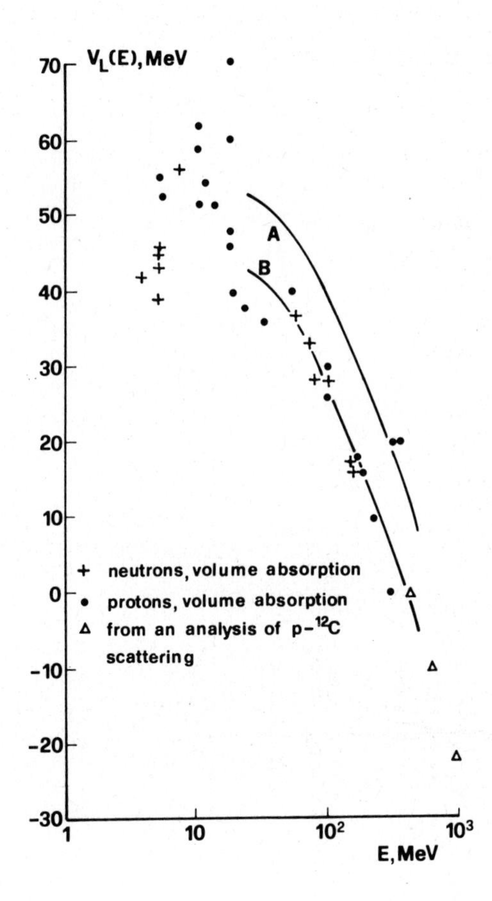

Fig.6. The real part of the equivalent local potential depth obtained from a Hartree-Fock calculation and a dispersion relation with the imaginary part shown in Fig. 5.
Curve A: unsubtracted dispersion relation; Curve B: subtracted dispersion relation. (From. Ref. 11).

about 10 MeV. Such a discrepancy may be interpreted as due to the high energy behaviour of the calculated imaginary part which is rather arbitrary. To avoid this difficulty the same calculation has been made by using a subtracted dispersion relation. In this way an additive parameter is introduced and is determined by normalizing the result at the energy of 400 MeV, where the empirical potential changes its sign. The agreement with the experimental data turns to be very good (Fig.6, curve B).

5.2 The relativistic treatments

As it concerns the second hypothesis made in the above calculations a completely different point of view can be assumed, as it has been made in the other works on the dispersion relation analysis.[3),4),7)]

The theoretical optical potential may be interpreted in a relativistic way. In fact, the concept of the theoretical optical potential is founded on the elimination of the anelastic channels in the formal description of the dynamics of the system. Such an elimination can be done also in a relativistic theory: the only difference is that also the anelastic channels due to particle production must be considered in addition to those of target excitation and breaking. This is made phenomenologically in a trivial way, when Eq. (2) is used in order to calculate the imaginary part of the empirical optical potential.

From a theoretical point of view the problem arises to see whether a second-quantization interaction between the projectile and the nucleus may change the general properties of the theoretical optical potential operator. I think that the analyticity properties, such as derived from a very general causal condition, would not change, while the spectral properties of the operator QHQ, and so the asymptotic behaviour of the theoretical optical potential, may be strongly affected.

A first implication that such a point of view has on the theoretical optical potential is the following.[4] If the imaginary part of $\mathcal{V}_L(E)$, Eq. (4), has the asymptotic behaviour (2), then the term $\mathcal{U}(E;\vec{r},\vec{r}')$ in Eq. (3) must contain a contribution which tends to be local as $E\to\infty$. This because the r.h.s. of Eq. (2) does not go to zero for $E\to\infty$, while, if $\mathcal{U}(E;\ s)$ were non local also for $E\to\infty$, one would have:

$$\text{Re}\ \mathcal{V}_L(E) \xrightarrow[E\to\infty]{} \mathcal{V}_1 \quad , \tag{14}$$

$$\text{Im}\ \mathcal{V}_L(E) \xrightarrow[E\to\infty]{} 0 \quad . \tag{15}$$

(The increasing oscillation of $\exp(i\text{Re}\tilde{\vec{k}}\cdot\vec{s})$ due to increasing $\text{Re}\tilde{k}$ would cancel the function $\mathcal{J}$ if the integral over $\vec{s}$ were over a not vanishing interval).

Such considerations lead to write the empirical optical potential in the following form:

$$\mathcal{V}_L(E) = \mathcal{V}_1 + \int \mathcal{V}_2(s)e^{i\tilde{\vec{k}}\cdot\vec{s}}d\vec{s} + \int \mathcal{U}(E;s)e^{i\tilde{\vec{k}}\cdot\vec{s}}d\vec{s} \quad , \tag{16}$$

where the term $\mathcal{U}(E;\ s)$ at the r.h.s. contains asymptotically a local E-dependent term whose imaginary part is

$$-\tfrac{1}{2}\hbar v\rho\gamma\bar{\sigma} \quad . \tag{17}$$

When the energy E becomes higher and higher, the wave length $\frac{2\pi}{\text{Re}\tilde{k}}$ becomes smaller and smaller, the contribution of the non-local terms to the r.h.s. of Eq. (16) becomes more and more depressed and when the energy is high enough only the contribution of the term going to local survives. This is the way in which Eq. (16) must be employed to analyse the empirical data. As, for practical calculations, it is not suitable to introduce a non locality depending on the energy, it is convenient to introduce a model which expresses the main features above discussed, i.e.:

1) $\mathcal{U}(E; s)$ contains a term going to local for $E\to\infty$;
2) Only such a term contributes to $\mathcal{V}_L(E)$ for E high enough;
3) The main contribution to the spurious energy dependence comes from the term $\mathcal{V}_2(s)$ (according to the Brueckner-Hartree-Fock calculation).

Such a model may be the following:

$$(18) \qquad \mathcal{V}_L(E) = \mathcal{V}_1 + \int \mathcal{V}_2(s) e^{i\tilde{\vec{k}}.\vec{s}} d\vec{s} + \mathcal{U}(E) \quad .$$

The term Re $\mathcal{U}(E)$ can be calculated by means of the dispersion relation by taking Im $\mathcal{V}_L(E)$ for its imaginary part (Fig. 2). A subtracted dispersion relation

$$(19) \qquad \mathrm{Re}\,\mathcal{U}(E) - \mathrm{Re}\,\mathcal{U}(E_o) = \frac{E-E_o}{\pi} \; P \int_o^\infty \frac{\mathrm{Im}\,\mathcal{V}_L(E')}{(E'-E)(E'-E_o)} dE'$$

must be used, and so only the slope of Re $\mathcal{U}(E)$ is meaningful.

If the ideas underlying the model (18) are realistic, one expects to find that, above a suitable energy, this slope agrees with that shown by the experimental data. It is found that this happens above 400 MeV and for this reason the subtracted dispersion integral (19) has been normalized to the highest energy experimental value (E_o = 970 MeV, $V_L(E_o)$ =-22.3 ± 4.3 MeV) (Fig. 1, full line). As the wave length at 400 MeV is about 1.2 fm, an estimate of the non-locality range of about 1 fm is indicated. This is a rather crucial point in order to test the ideas underlying the model (18). Unfortunately, the experimental points above 400 MeV are very few and affected by large errors; moreover some of them concern a light nucleus such a Carbon. They are the early results by Batty[12] on Carbon at 420, 610 and 970 MeV and those by Mc Manigal [13] and coworkers at 720 MeV on a range of nuclei. So more phenomenological analyses, in particular on heavy nuclei, in this energy range would be very important.

The term $\mathcal{V}_2(s)$ may be parametrized by assuming for it the Gaussian model (9), (10). A best fit is obtained with U = 96 MeV and β = 0.8 fm (Fig. 7, curve a).

It may be noted that, in this model, the function Re $\mathcal{U}(E)$ is given by the full line in the Fig. 1 (apart from an additive constant), whose derivative now expresses the dynamical energy dependence of $V_L(E)$. It may be interesting to compare this derivative with that calculated by Jeukenne, Lejeune and Mahaux in the energy range up 100 MeV. The sign is the same (positive, i.e. the dynamical contribution increases V_L in this energy range) and the average value obtained from the curve of Fig. 1 in the interval between 10 and 100 MeV is 0.07 while that obtained in the Brueckner-Hartree-Fock approximation is 0.12. It must be remarked that the latter value refers to the center of the nucleus, and it turns to be lowered when averaged on the nuclear density.

A dispersion relation analysis also based on the relativistic point of view was done by another group[7] nearly at the same time and independently. The starting point is the following formula for the equivalent optical potential:

$$\mathcal{V}_L(E) = \left(K_o(E) + \frac{m}{3\hbar^2}EK_2(E)\right)/\left(1 - \frac{m}{3\hbar^2}K_2(E)\right) \quad , \tag{20}$$

where:

$$K_n(E) = \int s^n \left(\mathcal{V}_2(s) + \mathcal{U}(E;s)\right) d\vec{s} \quad . \tag{21}$$

Eq. (20), which was given in an early paper by Feshbach,[14] is

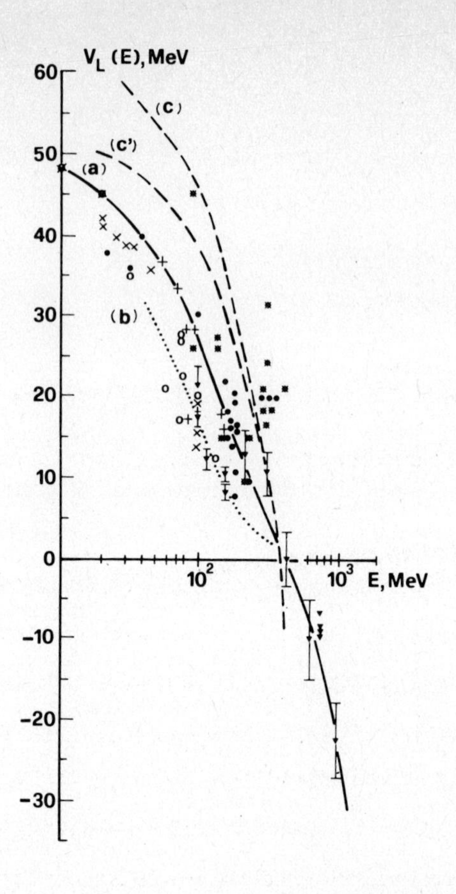

Fig. 7. Various predictions for the energy behaviour of the real part of the local empirical potential depth.

Curve (a): from a model based on the dispersion relation and non-locality of the theoretical optical potential; Curve (b): early calculation in terms of the nucleon-nucleon forward scattering amplitude; Curve (c): calculation in terms of the nucleon-nucleon forward scattering amplitude obtained from recent phase shift analyses; Curve (c'):off-shell correction to the curve (c). (From ref. 10).

simply the second order approximation to Eq. (4) in the development of $\mathcal{J}(E\ ;\ \tilde{k}^2)$:

$$\mathcal{J}(E,\tilde{k}^2) = \sum_{n=0}^{\infty} \frac{(-1)^n}{(2n+1)!}\ \tilde{k}^{2n}\ K_{2n}(E) \quad . \tag{22}$$

From Eqs. (20) and (21) a dispersion relation for the empirical optical potential is derived:

$$\mathrm{Re}\,\mathcal{V}_L(E) - \mathrm{Re}\,\mathcal{V}_L(E_o) = \frac{E-E_o}{\pi}\ P\int_0^{\infty} \frac{\mathrm{Im}\,\mathcal{V}_L(E')}{(E'-E)(E'-E_o)}\, dE' - E_o\alpha(E_o) + E\alpha(E) \quad . \tag{23}$$

Here $\alpha(E)$ is a rather complicated function which is parametrized by

$$\alpha(E) = A\ /\ (E + B) \quad . \tag{24}$$

With the values A = 78.3 MeV, B = 195.7 MeV, the experimental values are well reproduced (Fig. 8).

The approximations which lead to the dispersion relation (23) are essentially the formula (20) and the hypothesis that $K_2(E)$ may be neglected at high energies (E>1 GeV). The validity of these approximations is not quite clear. Eq. (20) is certainly a good low energy formula, but the convergence of the series (22) at high energies is very dubious, if the theoretical optical potential has a fixed non-locality. The authors assume that the theoretical optical potential becomes local at high energy, and this justifies both the use of Eq. (20) and the neglect of $K_2(E)$ above 1 GeV, but the non local term $\mathcal{V}_2(s)$ due to the antisymmetrization is independent of the energy. It is very different to say that at high energy the contribution to the empirical potential mainly comes from a term of the theoretical optical potential becoming local from saying that the whole theoretical optical potential becomes local. I think that the success of the simple parametrization (23) is due to the following reason: the development (22) may be ap-

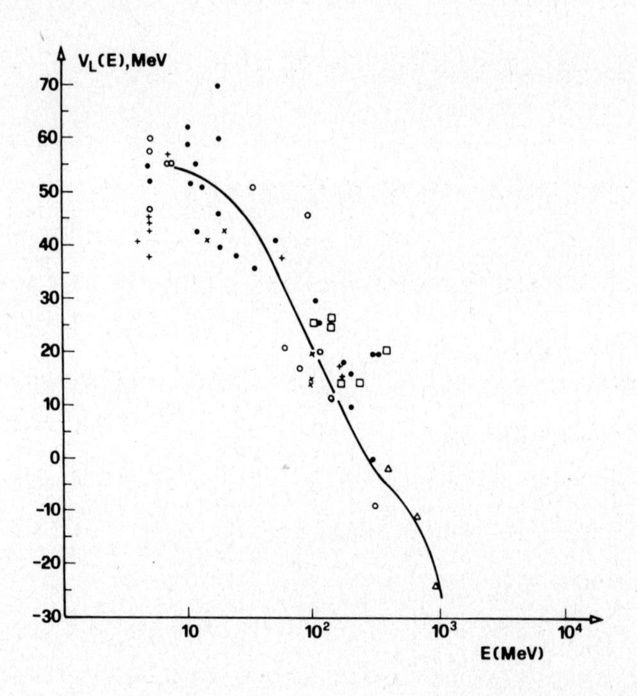

Fig. 8. The real part of the local empirical depth obtained from the formula (23) (From Ref. 7).

plied only to the part of the function $\mathcal{J}$ due to the term $\mathcal{U}(E\ ;\ s)$:

$$\int \mathcal{U}(E;s)e^{i\tilde{\vec{k}}.\vec{s}}d\vec{s} = \sum_{n=0}^{\infty} \frac{(-1)^n}{(2n+1)!}\ k^{2n}K_{2n}(E) \quad , \tag{25}$$

where now

$$K_{2n}(E) = \int s^n \mathcal{U}(E\,;s)\, d\vec{s} \quad . \tag{26}$$

The formula (20) turns to be somewhat modified:

$$\mathcal{V}_L(E) = \left(\mathcal{V}_1 + \int \mathcal{V}_2(s) e^{i\hat{\vec{k}}.\vec{s}} d\vec{s} + K_o(E) + \frac{m}{3\hbar^2} EK_2(E)\right) / \left(1 - \frac{m}{3\hbar^2} K_2(E)\right) . \tag{27}$$

The approximations made by the authors involve now only the term $\mathcal{U}(E\,;s)$ which has a lower non locality than $\mathcal{V}_2$ and contains a term which becomes local for $E \to \infty$. Thus the approximation of the truncation of the series (22) and the neglecting of $K_2(E)$ above 1 GeV are now well justified, and so the calculation of $K_o(E)$ by means of the dispersion relation. In this way the final formula (23) is again obtained, were now into the term parametrized by the function $\alpha(E)$ also the term $\int \mathcal{V}_2(s)\exp(i\hat{\vec{k}}.\vec{s})d\vec{s}$ must be included. In fact, the function $\alpha(E)$ is a low energy correction to the dispersion relation, as the term $\int \mathcal{V}_2(s)\exp(i\hat{\vec{k}}\cdot\vec{s})d\vec{s}$ is.

In conclusion : all the analyses here reviewed, either they assume the non-relativistic point of view or the relativistic one, agree in finding a dominant spurious energy dependence for the real part of the empirical optical potential below 200 MeV. This agrees with the very recent calculation in Brueckner-Hartree-Fock approximation[6)] and indicates a range of non-locality of the theoretical optical potential of about 1 fm. The results of these treatments are very similar, in spite of their important conceptual difference, because the dispersion integral, at low energies, is quite insensitive to the high energy behaviour of the imaginary part, expecially when it is used in a subtracted form. Of course, this is no longer true at a higher energy, (above 400 MeV) where a better experimental information would be very important.

The relativistic treatments cover the whole region where experimental values are known, i.e. the region from about 10 MeV to 1 GeV and indicate that the spurious energy dependence ceases to dominate in the region from 300 to 400 MeV, after that the energy dependence becomes completely dynamical.

6. Comparison with the calculation in terms of nucleon-nucleon scattering amplitude

The calculations reviewed up to this point are based on the expression of the theoretical optical potential in terms of the spectrum of the operator QHQ given in the introductory paper. We compare them now with calculations directly made on the formula which expresses the theoretical optical potential in terms of the forward nucleon-nucleon scattering amplitude:

$$\langle \vec{k}' | \mathcal{V} | \vec{k} \rangle = \langle \vec{k}', \vec{k}' - \vec{k} | t(E) | \vec{k}, 0 \rangle \hat{\rho} (\vec{k} - \vec{k}') \quad . \tag{28}$$

About the use of such an expression, there is, in my opinion, some confusion. It is well known the early result by Riesenfeld and Watson[9], who calculated the optical potential in terms of the nucleon-nucleon phase-shifts up to about 300 MeV, obtaining a good agreement with the empirical data (Fig. 7, curve b). But the question is [10]: do the calculations of this type concern a local approximation to the theoretical optical potential, or the local transform of the theoretical optical potential in the sense of Eq. (3)?

In the first case a striking contradiction would arise with the dispersion relation analyses above described, which imply a large non-locality of the theoretical optical potential. The answer is that such calculations concern the local transform of the theoretical optical potential, and thus their agreement with the empirical data does not at all imply that the theoretical optical potential is local, but rather that the hypotheses underlying the expression of the theoretical optical potential in terms of the nucleon-nucleon scattering amplitude (mainly the simple scattering and impulse approximations) are correct. This can be seen by starting from the equation

$$\langle\vec{k}'|\mathcal{V}|\vec{k}\rangle = (2\pi)^3 \rho(0)\,\delta(\vec{k}-\vec{k}')\langle\vec{k}0|t(E)|\vec{k}0\rangle \quad , \tag{29}$$

which is Eq. (28) in the limit case of infinite nuclear matter. It must be emphasized that there the matrix $\langle\vec{k},0|t(E)|\vec{k},0\rangle$ is not on the energy shell: E is the energy of the projectile, but $\vec{k}$ is the label of the matrix element in the momentum representation. It can easily be seen that the equivalent local potential of the non local potential $\mathrm{Re}\langle\vec{k}'|\mathcal{V}|\vec{k}\rangle$ is

$$(2\pi)^3 \rho(0)\ \mathrm{Re}\langle\hat{\vec{k}}0|t(E)|\hat{\vec{k}}0\rangle \quad , \tag{30}$$

where $\hat{\vec{k}}$ is the momentum under this potential:

$$\left((\hat{k}^2)/(2m)\right) + (2\pi)^3 \rho(0)\ \mathrm{Re}\langle\vec{k}0|t(E)|\vec{k}0\rangle = E \quad . \tag{31}$$

Such a potential can be identified with the real part of $\mathcal{V}_L$

$$\mathrm{Re}\,\mathcal{V}_L(E) \equiv (2\pi)^3 \rho(0)\,\mathrm{Re}\langle\hat{\vec{k}}0|t(E)|\hat{\vec{k}}0\rangle \quad , \tag{32}$$

and thus

$$\left((\hat{k}^2)/(2m)\right) + \mathrm{Re}\,\mathcal{V}_L(E) = E \quad . \tag{33}$$

For $E \gg \mathrm{Re}\,\mathcal{V}_L(E)$, the t-matrix in Eq. (30) is well approximated by that on the energy shell. Eqs. (30) and (33) clarify the meaning of the calculations of the empirical potential in terms of nucleon-nucleon phase shifts. In order to calculate the empirical optical potential depth the following rule can be used:

At low energy, if one neglects the E-dependence of the t-matrix, the optical potential $\mathrm{Re}\,\mathcal{V}_L(E)$ is proportional to the forward nucleon-nucleon scattering amplitude calculated at the energy $E - \mathrm{Re}\,\mathcal{V}_L(E)$, that is

(34) $$-\mathrm{Re}\,\mathcal{V}_L(E - C\,\mathrm{Re}\,f(E)) = C\,\mathrm{Re}\,f(E) \quad ,$$

where f(E) is the forward nucleon-nucleon scattering amplitude and

(35) $$C = (2\pi\hbar^2/m)\,\rho(0) \quad .$$

At high energies, $E \gg \mathrm{Re}\,\mathcal{V}_L(E)$, we have simply:

(36) $$-\mathrm{Re}\,\mathcal{V}_L(E) = C\,\mathrm{Re}\,f(E) \quad ,$$

with C given by Eq. (35) and, in the relativistic region, by:

(37) $$C = \left[(2\pi\hbar^2 c^2)/(E_t)\right]\rho(0) \quad ,$$

where E_t is the total relativistic energy of the projectile in the laboratory.

According to this rule, the real potential depth $V_L(E)$ has been calculated by using the more recent data on nucleon-nucleon phase shifts and on the nucleon-nucleon forward scattering amplitude. The results are shown by the curve (c') in Fig. 7 which does not contain free parameters and gives an information how the simple scattering and the impulse approximations hold at various energies.

References

1) H.Feshbach, Ann. Phys. 5, 357 (1958); 19, 287 (1962).

2) R.Lipperheide, Nucl. Phys. 89, 97 (1966).

3) G.Passatore, Nucl. Phys. A95, 694 (1967).

4) G.Passatore, Nucl. Phys. A110, 91 (1968).

5) R.Lipperheide and A.K.Schmidt, Nucl.Phys. A112, 65 (1968).

6) H.Fiedeldey and C.A.Engelbrecht, Nucl.Phys. A128, 673 (1969).

7) I.Ahmad and M.Z.Rahman Khan, Nucl.Phys. A132, 213 (1969).

8) J.P.Jeukenne, A.Lejeune and C.Mahaux, to be published.

9) W.B.Riesenfeld and K.M.Watson, Phys. Rev. 102, 1157 (1956).

10) G.Passatore, Nucl. Phys. A248, 509 (1975).

11) G.Eckart and M.K.Weigel, to be published.

12) C.J.Batty, Nucl. Phys. 23, 562 (1961).

13) P.G.Mc Manigal, R.D.Eandi, S.N.Kaplan and B.J.Moyer, Phys. Rev. 137, B620 (1965).

14) H.Feshbach, Ann.Rev. Nucl. Science 8, 49 (1958).

THEORETICAL INVESTIGATIONS OF THE OPTICAL-MODEL POTENTIAL†

J.-P. JEUKENNE, A. LEJEUNE and C. MAHAUX

Institut de Physique, Université de Liège

4000 Liège I, Belgium

Abstract. We present a critical survey of recent theoretical calculations of the complex optical-model potential. We argue that calculations with the smallest number of parameters carry the most meaningful information, and we organize the discussion according to this criterion.

1. Introduction

A theoretical study of the optical model should pursue three goals, which are somewhat interrelated:

a) Explain why the optical model is successful, despite the strong character of the nucleon-nucleon interaction. This is mainly studied in the frame of the many-body problem, as discussed in our accompanying paper.[1)]

b) Show that there exists a definition of the optical-model potential (OMP) which is in keeping with the use that is made of the optical-model phase shifts and also of the optical-model wave functions, which is a more delicate point. Here, we adhere to the definition given in Ref. 2: we identify the OMP with the mass operator.

c) This paper is devoted to a third problem, namely to the constraints that can be imposed on theoretical grounds on the parametric form of the OMP, and also to the information that the theory can give

---

† presented by C. MAHAUX

on some components of the OMP that are not easily accessible experimentally. For instance, one can try to calculate the scalar and tensor spin components or the scalar and tensor spin-isospin components of the OMP.[3] Here, we shall only deal with the central part of the OMP.

This central part $M(\vec{r},\vec{r}\,';E)$ of the mass operator is still quite a complicated function since it is nonlocal, energy-dependent and complex. One task for the theorist is to show when and how one can replace this complicated operator by an "equivalent" one, which would for instance be independent of energy (but still nonlocal) or local (but still energy-dependent). This problem is discussed for instance in Ref. 4 in the case of nuclear matter and in Refs. 5 to 8 in the case of finite nuclei. It appears that very little experimental information is available on the "true" nonlocality and on the "true" energy dependence of $M(\vec{r},\vec{r}\,';E)$. This is because these two factors cannot be disentangled empirically : this is thus a nice example where a theoretical investigation is useful. Here, we shall not discuss the information provided by the dispersion relation that relates the real and the imaginary parts of the OMP (for a review, see Ref. 9),since we presume that it is described elsewhere in these Proceedings.

A more delicate role of the theory is to give a guide-line on the parametric form of the OMP which should be used in empirical analyses of the data if one wants them to yield physically meaningful parameters. This is quite a difficult task, since the constraints derived from the theory are meaningful only if they are based on a reasonable approximation concerning the reaction dynamics and the nuclear interaction, and if it involves at most very few (ideally no !) adjusted parameter.

We make no attempt at completeness: the choice of the calculations that are discussed below is mainly based on exemplarity and recentness. Moreover, our appraisal must be taken with due caution, since it is unavoidably influenced by unconscious prejudices. The various theoretical approaches are so numerous and diversified that it is difficult to present them in any logical manner. In keeping with the remarks made above, we have grouped the calculations according to the number of adjustable parameters that they involve, rather than according to the underlying theoretical techniques, or energy domain. The real part of the OMP is discussed in Sect. 2, and its imaginary part in Sect. 3.

2. Calculations of the real part of the OMP

2a. Realistic nucleon-nucleon interactions

Realistic nucleon-nucleon forces have been used in the framework of Brueckner's theory (see Ref. 1), of the Martin-Schwinger set of coupled equations [10,11] and of the multiple scattering theory (see below).The Brueckner-Hartree-Fock (BHF) approximation can be used to distinguish the true nonlocality from the true energy dependence of the OMP; it can be extended to the spin-spin, isospin, spin-isospin components of the OMP; it appears to yield good agreement with empirical values of the OMP.[1] The main problem is to estimate the accuracy of the BHF and local density approximations. This approach can be used up to the energy domain where the concept of a nucleon-nucleon potential looses its meaning, i.e. up to about 300 MeV. Beyond this energy, one can use the impulse approximation, which is the leading term of the multiple scattering series. A detailed comparison between the BHF and the multiple scattering series in the domain 100-300 MeV would be of interest, since there both are expected to be fairly accurate. The multiple scattering series can cope with centre-of-mass and with Fermi motion; it has, however, some problems with the Pauli principle (antisymmetrization between the projectile and the target nucleons) and with off-shell effects. The latter have been investigated e.g. by Lerner and Redish.[12] These authors study a three-body model(projectile, one active bound nucleon and an inert core) and calculate the scattering of 65 MeV protons by ^{16}O , ^{17}O and ^{18}O . They use Reid's soft core interaction and find that off-shell effects are rather large, as expected at this low energy. The calculated result is in fair agreement with the empirical OMP only at the nuclear surface; it is considerably too deep in the nuclear interior. Nevertheless, it yields good cross sections if a phenomenological imaginary part is added. Off-shell effects become small above 120MeV. This, however, does not mean that the impulse approximation is already good at these energies.[13]

The calculations described above[1,12,14] are based on a strong nuclear force. Weaker but still "realistic" nucleon-nucleon interactions have been constructed. For instance, Tabakin's separable interaction[15] has been used by Rook[16] in the frame of Brueckner's theory and by Mackellar, Reading and Kerman[17] for the study of the scattering of low energy (a few MeV) neutrons by ^{16}O in second-order perturbation theory. It would be instructive to use still weaker but nevertheless realistic forces which have more recently been constructed.[18,19]

2b. Physical effective interactions

We coin "physical effective" an interaction that is not adjusted to the OMP to be calculated, but which is taken a priori from other nuclear properties. One example is the Skyrme three-body contact force whose parameters have been fitted to the bound state properties of nuclei. Dover and Van Giai [20,21] have used the Skyrme interaction for a

Hartree-Fock calculation of the OMP. Their calculation is self-consistent in the sense that they compute the nuclear density from the Skyrme interaction; these authors also investigate the spin-orbit and symmetry components of the OMP. As in all calculations which involve an effective interaction, one cannot disentangle nonlocality from energy dependence; consequently, this obliterates the significance of the Coulomb correction (see Ref. 1) computed in Ref. 21. The agreement between the theoretical and the empirical values of the OMP is fair at low energy (below 30 MeV) but the calculated energy dependence is too large. This is shown in Fig. 1, taken from Ref. 23. The theoretical diffuseness (≃ 0.55 fm) is quite small; this is incidentally also a feature that we observe in the many-body calculation described in Ref. 1.The Skyrme force has also been recently used by Manweiler.[7] Note that these studies only deal with the real part of the OMP.

Slanina and McManus[24] have calculated the real part of the OMP from the folding formula

$$(1) \qquad V(r) = \int t(\vec{r} - \vec{r}\,')\rho(\vec{r}\,')\, d^3r' \quad .$$

They take the nuclear density $\rho(r)$ from experiment (as in Ref. 1) and adopt for the effective interaction $t(\vec{r} - \vec{r}\,')$ either an interaction that fits nucleon-nucleon phase shifts at low energy, or density-dependent forces due respectively to Kuo and Brown and to A.M. Green. They also estimate the correction to Eq. (1) that arises from the exchange (Fock) term. The latter introduces a nonlocal contribution whose size has also been investigated by other authors [25,26] and appears somewhat too small to account for the empirical nonlocality.[26] This may be due to the fact that the effective interaction itself should be nonlocal (and energy-dependent). The interest of this type of study is main

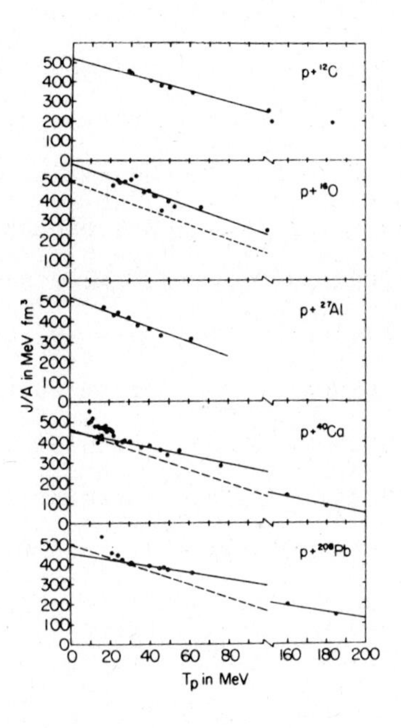

Fig. 1. The dots represent empirical values of the volume integral per nucleon of the real part of the OMP. The full straight line is a least square fit to the dots that lie above 25 MeV. The dashed straight lines are the values calculated by Dover and Van Giai.[21,23)

ly to show that the effective interaction used in nuclear structure calculations has about the right strength to reproduce the empirical OMP when Eq. (1) is used. A more detailed discussion can be found in Ref. 27.

2c. Effective interactions

Following the pioneering work of Greenlees and collaborators[28] it has become fashionable to fit the experimental data with an OMP calculated from the folding formula (1) (plus possibly the exchange term), with a phenomenological effective interaction t. The latter contains adjustable parameters and is only loosely related to the nucleon-nucleon interaction. It appears quite dangerous to consider this "reformulation" of the OMP as a physically justified constraint, in view of the phenomenological nature of the effective interaction. For instance, it is not reliable to investigate in this way the density distribution $\rho(r)$: if a density-independent effective interaction is used in Eq.(1), the density distribution ρ which reproduces the empirical OMP can be quite incorrect. This has been nicely demonstrated by Myers[29] (see also Ref. 30). Thus, the reduction of the number of parameters when Eq. (1) is used with an effective interaction t may lead to unphysical constraints. An exhaustive list of references concerning this "reformulated" OMP can be found in Ref. 27.

Effective interactions have also been used recently by Manweiler[7] and by Giannini and Ricco [31] for constructing the OMP. Their approaches share several features. In particular, they try to construct a potential well which would reproduce not only the scattering but also the bound state data (single-particle energies, density distribution).

Manweiler [7] extends to the continuum a model which had previously been used by Elton, Webb and Barrett [22] for bound states. It consists in a Schrödinger equation with a nonlocal interaction of the Perey-Buck form:[5]

$$(2) \qquad V(\vec{r},\vec{r}\,') = U\left(\frac{\vec{r} + \vec{r}\,'}{2}\right) \exp\left(- \left(\frac{\vec{r} - \vec{r}\,'}{\beta}\right)^2\right) ,$$

where U is assumed to have a Woods-Saxon shape. The various parameters are then adjusted to the bound and scattering data.
Manweiler[7] gives a qualitative justification of Eq. (2) based on many-body theory. Hence, his constrained parametrization may have some meaning, in the sense that it may yield parameters that one could in principle calculate from a nuclear matter approach with a realistic interaction. Figure 2 shows the radial dependence of U in Eq. (2) in

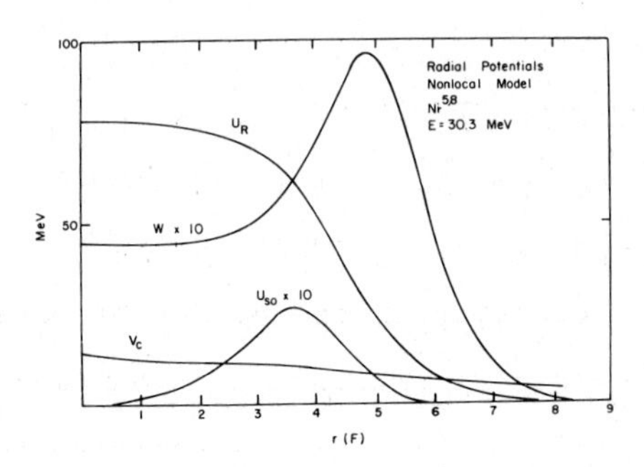

Fig. 2. Radial dependence of U(r) (see Eq. (2)) as determined by Manweiler [7] in the case ^{58}Ni at 30.3 MeV. The labels U_R, U_{SO} and W refer to the central, spin-orbit and imaginary parts of the OMP, respectively; V_C is the Coulomb field.

the case of ^{58}Ni at 30.3 MeV: U_R is the real part of U, V_C the Coulomb field, U_{SO} the spin-orbit component and W the imaginary part of the potential. The observed agreement between empirical and calculated values is in our opinion not very meaningful in view of the number of adjusted parameters. It appears to us that the main interest of this work is to show that it is possible to fit bound and low energy (<50 MeV) scattering data with a nonlocal and energy-independent OMP whose parametric form appears rather plausible in the framework of many-body theory.

The last sentence also applies to the recent work by Giannini and Ricco,[8)] provided that one replaces "many-body" by "multiple scattering". The work described in Ref. 8 pays more attention to the centre-of-mass motion. However, it involves some approximations which are of dubious validity when extrapolated from the high energy to the low energy or *a fortiori* negative energy domain. We have for instance in mind the impulse approximation and the omission of the antisymmetrization between the projectile and the target nucleons. The latter is believed to be responsible for a fraction of the observed energy dependence of the OMP. Hence, the adjusted parameters of the effective interaction somehow have to compensate for these approximations; this weakens their physical meaning. The equivalent local potential (ELP) is found to have a geometry (depth, radius, diffuseness) which depends on energy. The full dots in Fig. 3 represent empirical potential depths (of the ELP), the full curve is the parametrization of Giannini and Ricco;[8)] the long dashes show the depth of the ELP that we calculated from Reid's hard core nucleon-nucleon interaction.[1)] Note that the full and dashed curves are not directly comparable since they correspond to different geometries.

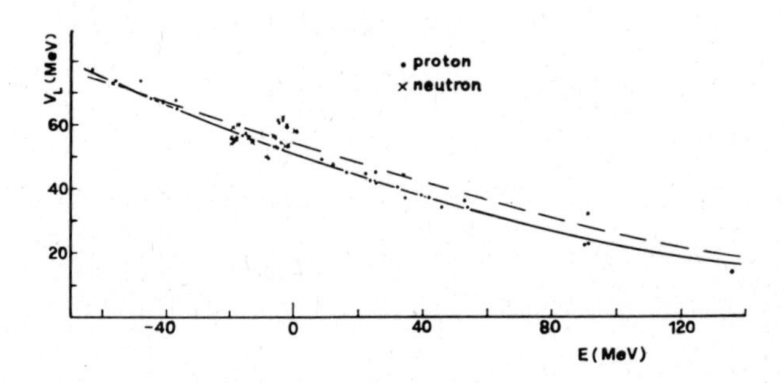

Fig. 3. The dots represent the depth of the equivalent local potential versus energy, with the geometry of Giannini and Ricco;[31] the full curve shows their parametric fit. The dashes give the depth of the OMP calculated from Brueckner's theory.[4] Note that the two curves correspond to different geometries.

3. Calculations of the imaginary part of the OMP

3a. Realistic nucleon-nucleon interaction

It is only in nuclear matter that realistic nucleon-nucleon interactions have been used to calculate the imaginary part of the OMP. A local density approximation (LDA) is then used to calculate the OMP in a finite nucleus. We note that the justification of this LDA is more questionable in the case of the imaginary part of the OMP, which is expected to be more sensitive to nuclear structure [31] and to shell effects [32,33] than the real part of the OMP. Nevertheless, it appears that this type of approach is rather successful.[1] However, deviations should be expected in individual cases: the many-body theory can only give the "global" parametric form of the imaginary part of the OMP.

Nuclear matter calculations with realistic interactions have been based either on the Martin-Schwinger-Puff theory[10,11,34,35] or on Brueckner's approach.[1,36] The first method is not satisfactory at low energy, at least in the form that was used until now. The second approach has met with more success.[1] It can be extended to the isospin, spin-spin and spin-isospin components of the OMP.[3,37] It is found [4] that, at low energy, the energy dependence of the empirical absorption is dominated by the dynamical (true) energy dependence of the mass operator. This is in keeping with the frivolous model of Clementel and Villi [38] and also substantiates assumptions that had been made in connection with the dispersion relation that relates the real to the imaginary part of the OMP.[9]

3b. Physical effective interactions

As in Sect. 2b, we coin "physical effective" those effective interactions which are not adjusted to the empirical OMP but are rather taken from other sources, e.g. from other nuclear structure calculations. Approaches based on this kind of forces are of particular interest in the case of the imaginary part of the OMP, in view of the possible importance of nuclear structure effects.

The absorptive part of the OMP has two main origins, namely: (a) The existence of open inelastic and reaction channels; (b) The average over compound nuclear resonances: part of the incoming flux feeds the compound nucleus. To our knowledge, Cugnon [39] is the only author who included both effects explicitely,in the case p + ^{39}K . The compound states (of ^{40}Ca) are described as one particle-one hole excitations and the excited states of ^{39} K as one-hole states. The latter approximation may underestimate the role of direct transitions to inelastic channels, since it omits collectivity effects. With this proviso in mind, Cugnon [39] finds that effects (a) and (b) are comparable, but the calculated absorption is somewhat too small.

The compound nucleus contribution to the absorptive part was calculated in the case of the target ^{58}Ni in Refs. 40, and 41, and in the case of ^{208}Pb in Refs. 42 to 44. These works are limited to low or negative energy. The compound states are described in a particle-vibration coupling model. Since the latter includes only a few degrees of freedom, the calculated imaginary part has a nonmonotonic energy dependence (resonances) and an awkward nonlocality. An example is shown in Fig. 4 (from Ref. 44).We see that the results cannot be readily compa-

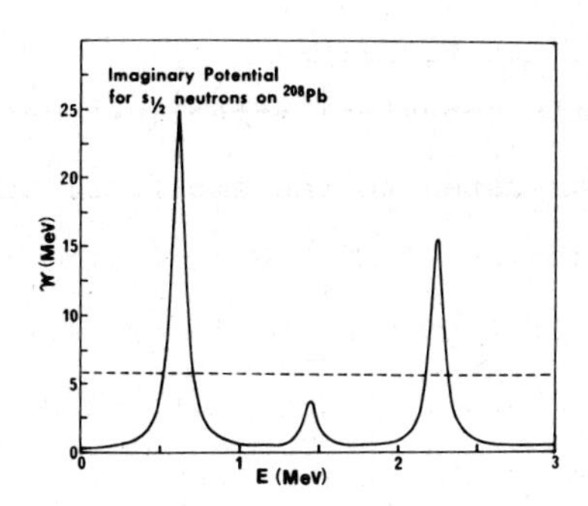

Fig. 4. The full curve shows the imaginary part of the OMP as calculated in Ref. 44, in the case of ^{208}Pb. The dashed line is the empirical value (from Ref. 44).

red with empirical OMP; the main interest of these calculations rather lies in the interpretation of the bumps that may be observed in excitation cross sections or in strength function distributions. We note, however, that it is usually believed that compound nuclear formation is the main source of volume absorption,[45] so that these calculations may be useful in this respect, if extrapolated to higher energy.

We now turn to the absorption which is due to direct transitions into inelastic channels. Rao, Reeves and Satchler [45,46] studied the scattering of 30 MeV protons on ^{40}Ca and ^{208}Pb. Since the target excited states are derived from the (collective) deformed potential model, this is a macroscopic calculation. The coupling strengths are taken from inelastic scattering measurements. If sum rules are used to estimate the effect of unobserved excited states and if the (important, see Ref. 47)contribution of pick-up channels is roughly included, good absorption cross sections are obtained. However, the angular distributions are in poor agreement with the experimental data.

A detailed microscopic calculation of the imaginary part due to inelastic channels was recently completed by Vinh Mau and Bouyssy,[48] in the case of p + ^{40}Ca. The intermediate (open) channels are described by the random phase approximation (RPA). Thus, collectivity effects are included microscopically. The Pauli principle is carefully taken into account. The main approximations are the neglect of compound nuclear formation, the choice of a zero-range "physical effective" force, and the use of plane wave intermediate states. Figure 5 (from Ref.48) shows the contributions of various target excited states with negative parity to the imaginary part of the mass operator $M(\vec{r},\vec{r}\,';E)$, for $s = |\vec{r} - \vec{r}\,'| = 0.02$ fm and E = 14 MeV; one has $R \simeq \frac{1}{2}(r + r')$. This kind of study thus enables one to identify the contribution of the va-

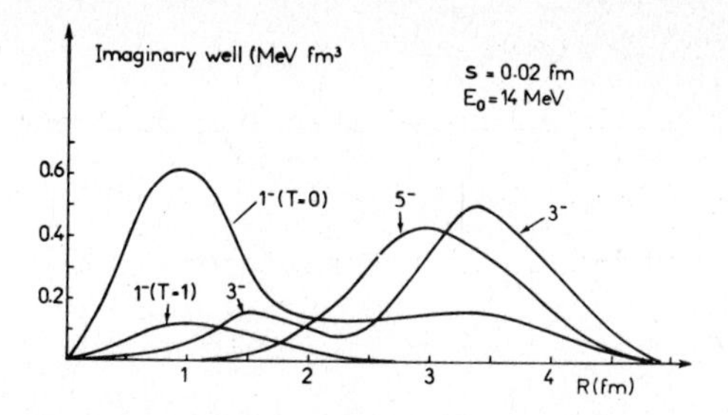

Fig. 5. Contributions of various inelastic channels to the imaginary part of the OMP in ^{40}Ca at 14 MeV, as calculated by Vinh Mau and Bouyssy.[48)]

rious channels, and to disentangle the true nonlocality from the true energy dependence.

In Sect. 2a, we listed a few nuclear matter calculations that are based on a realistic nucleon-nucleon potential. A simplified version of this "Fermi gas" approach is the frivolous model [38)] where, essentially, the role of the effective potential is replaced by the forward nucleon-nucleon amplitude. This neglects off-shell and binding effects. It is found that at least one parameter must be adjusted, [49,50)] but the resulting fits are then quite good.

3c. Effective interactions

Manweiler [7)] and Giannini and Ricco [8)] have included a nonlocal absorptive part in their models (see Sect. 2c). As expected from the frivolous model, one must in addition [51)] include a suitable explicit energy dependence which takes account of the reduction of the phase space when the energy is decreased. This was not found necessary by Perey and Buck[5)] in their pionneering analysis. The latter phenomenological observation was at the origin of some misunderstanding concerning the nonlocality of the imaginary part of the OMP.

4. Conclusions

In the present review, we essentially adopted the point of view that a calculation of the OMP with adjustable parameters is only of limited interest. Indeed the variability of the parameters permits one to mock up omitted processes in a manner that can be misleading in the sense that it corresponds to imposing unphysical constraints on the calculated OMP. This attitude is shared by Lerner and Redish[12)] who stated that "the utility of a theory of the optical model can only be investigated by calculating the first term directly, without free parameters". This does not imply that one should start from (almost) "first principles", i.e. from a realistic nucleon-nucleon interaction (Sects. 2a and 3a). Indeed, one can also use the vast amount of information that is now available both on effective interactions and on the theory of nuclear low excited states. This enables one to give reasonable constraints on the OMP at low energy, by extrapolation of nuclear structure calculations (Sects. 2b and 3b). Of course, some details are then lost (e.g. true nonlocality) which can be of practical significance. Calculations with adjustable interactions (Sects. 2c and 3c) can also be of interest if they are derived from a reasonable approximation scheme for the theoretical OMP: the rendering of the adjusted parameters from a many-body calculation is then a useful challenge to the theorist and the constraints may be meaningful, although this is very hard to judge.

References

1) J.-P.Jeukenne, A.Lejeune and C.Mahaux, Many-Body Theory of the Optical-Model Potential, in these Proceedings.

2) J.S.Bell, in Lectures on the Many-Body Problem, edited by E.R. Caianiello (Academic Press, N.Y., 1962) p. 91.

3) J.Dabrowski and P.Haensel, Can.Journ.Phys. 52, 1768 (1973).

4) J.-P.Jeukenne, A.Lejeune and C.Mahaux, Phys. Reports (1976).

5) F.Perey and B.Buck, Nucl. Phys. 32, 353 (1962).

6) E.H.Canfield, H.J.Amster, R.G.Kasper and H.Mark, Nucl. Phys. A169, 385 (1971).

7) R.W.Manweiler, Nucl. Phys. A240, 373 (1975).

8) M.M.Giannini and G.Ricco, preprint Genova, November 1975.

9) M.Bertero and G.Passatore, Z.Naturforsch. 28a, 519 (1973).

10) H.Gall and M.R.Weigel, Z.Physik A276, 45 (1976).

11) C.Marville, preprint (Liège, 1976).

12) G.M.Lerner and E.F.Redish, Nucl. Phys. A193, 565 (1972).

13) R.C.Johnson and D.C.Martin, Nucl. Phys. A192, 496 (1972).

14) H.R.Kidwai and J.R.Rook, Nucl. Phys. A169, 417 (1971).

15) F.Tabakin, Ann.Phys. (N.Y.) 30, 51 (1964).

16) J.R.Rook, Nucl.Phys. A222, 596 (1974).

17) A.D.MacKellar, J.F.Reading and A.K.Kerman, Phys.Rev. C3, 460 (1971).

18) D.Gogny, P.Pires and R.de Tourreil, Phys.Letters 32B, 591 (1970).

19) R.de Tourreil and D.W.L.Sprung, Nucl.Phys. A201, 193 (1973).

20) C.B.Dover and Nguyen van Giai, Nucl.Phys. A177, 559 (1971).

21) C.Dover and Nguyen van Giai, Nucl.Phys. A190, 373 (1972).

22) L.R.B.Elton, S.J.Webb and R.C.Barrett, Phys.Rev.Letters 24, 145 (1970).

23) W.T.H.Van Oers and H.Haw, Phys. Letters 45B, 227 (1973).

24) D.Slanina and H.McManus, Nucl.Phys. 116, 271 (1968).

25) L.W.Owen and G.R.Satchler, Phys. Rev. Letters 25, 1720 (1970).

26) W.G.Love and L.W.Owen, Nucl.Phys. A239, 74 (1975).

27) B.Sinha, Phys. Reports 20C, 1 (1975).

28) G.W.Greenlees, G.J.Pyle and Y.C.Tang, Phys. Rev. 171, 1115 (1968).

29) W.D.Myers, Nucl.Phys. A204, 465 (1973).

30) D.K.Srivastava, N.K.Ganguly and P.E.Hodgson, Phys. Letters 51B 439 (1974).

31) G.E.Brown, Comments Nucl.Part.Phys. 4, 75 (1970).

32) A.Sugie, Phys. Rev. Letters 4, 286 (1960).

33) D.Gross, Z.für Physik 207, 251 (1967).

34) A.S.Reiner (Rinat), Phys. Rev. 133, B1105 (1964).

35) Q. Ho-Kim and F.C.Khanna, Ann.Phys. (N.Y.) 86, 233 (1974).

36) G.L.Shaw, Ann. Phys. (N.Y.) 8, 509 (1959).

37) P.Haensel, Nucl. Phys. A245, 29 (1975).

38) E.Clementel and C.Villi, Nuovo Cim. 2, 176 (1955).

39) J.Cugnon, Nucl. Phys. A208 , 333 (1973).

40) T.F.O'Dwyer, M.Kawai and G.E.Brown, Phys. Letters 41B, 259 (1972).

41) N.Azziz and R.Mendez-Placido, Phys.Rev. C8, 1849 (1973).

42) A.Lev, W.P.Beres and M.Divadeenam, Phys. Rev.Letters 31, 555 (1973).

43) A.Lev, W.P.Beres and M.Divadeenam, Phys. Rev. C9, 2416 (1974).

44) A.Lev. and W.P.Beres, Phys.Letters 58B, 363 (1975).

45) G.R.Satchler, International Symposium on Correlations in Nuclei (Balatonfüred, Hungary, Sept. 3-8, 1973), J.Németh, ed., Technoinform, Budapest (1974).

46) C.L.Rao, M.Reeves III and G.R.Satchler, Nucl. Phys. A207, 182 (1973).

47) R.S.Mackintosh, Nucl.Phys. A230, 195 (1974).

48) N.Vinh Mau and A.Bouyssy, Nucl. Phys. A257, 189 (1976).

49) B.Sinha and F.Duggan, Phys. Letters 47B, 389 (1973).

50) B.Sinha and F.Duggan, Nucl.Phys. A226, 31 (1974).

51) H.Fiedeldey and C.A.Engelbrecht, Nucl.Phys. A128, 673 (1968).